Gea Olbricht

Reproduction and growth of elephant shrews or sengis (Macroscelidea)

Gea Olbricht

Reproduction and growth of elephant shrews or sengis (Macroscelidea)

Some aspects of the reproduction of sengis in general and the post-natal growth of the short-eared sengi (Macroscelides proboscideus) in particular

Südwestdeutscher Verlag für Hochschulschriften

Impressum / Imprint
Bibliografische Information der Deutschen Nationalbibliothek: Die Deutsche Nationalbibliothek verzeichnet diese Publikation in der Deutschen Nationalbibliografie; detaillierte bibliografische Daten sind im Internet über http://dnb.d-nb.de abrufbar.
Alle in diesem Buch genannten Marken und Produktnamen unterliegen warenzeichen-, marken- oder patentrechtlichem Schutz bzw. sind Warenzeichen oder eingetragene Warenzeichen der jeweiligen Inhaber. Die Wiedergabe von Marken, Produktnamen, Gebrauchsnamen, Handelsnamen, Warenbezeichnungen u.s.w. in diesem Werk berechtigt auch ohne besondere Kennzeichnung nicht zu der Annahme, dass solche Namen im Sinne der Warenzeichen- und Markenschutzgesetzgebung als frei zu betrachten wären und daher von jedermann benutzt werden dürften.

Bibliographic information published by the Deutsche Nationalbibliothek: The Deutsche Nationalbibliothek lists this publication in the Deutsche Nationalbibliografie; detailed bibliographic data are available in the Internet at http://dnb.d-nb.de.
Any brand names and product names mentioned in this book are subject to trademark, brand or patent protection and are trademarks or registered trademarks of their respective holders. The use of brand names, product names, common names, trade names, product descriptions etc. even without a particular marking in this work is in no way to be construed to mean that such names may be regarded as unrestricted in respect of trademark and brand protection legislation and could thus be used by anyone.

Verlag / Publisher:
Südwestdeutscher Verlag für Hochschulschriften
ist ein Imprint der / is a trademark of
OmniScriptum GmbH & Co. KG
Heinrich-Böcking-Str. 6-8, 66121 Saarbrücken, Deutschland / Germany
Email: info@svh-verlag.de

Herstellung: siehe letzte Seite /
Printed at: see last page
ISBN: 978-3-8381-1180-3

Zugl. / Approved by: Duisburg-Essen, Universität, Diss., 2009

Copyright © 2009 OmniScriptum GmbH & Co. KG
Alle Rechte vorbehalten. / All rights reserved. Saarbrücken 2009

When we try to pick anything for itself,

then it turns out that it is linked to everything else in the universe.

John Muir

Was wir wissen, ist ein Tropfen;

was wir nicht wissen, ein Ozean.

Isaac Newton

Es ist nicht schwer zu komponieren.

Aber es ist fabelhaft schwer,

die überflüssigen Noten unter den Tisch fallen zu lassen.

Johannes Brahms

Meiner Familie gewidmet,

Dr. Alexander Sliwa mit Leona, Feline und Olivia

ACKNOWLEDGMENTS

Six years came and went in the blink of an eye. Through it all, I've had a great deal of fun and it is a great pleasure for me to acknowledge all those who've helped me in this endeavour. In 2002 I approached Professor Hynek Burda of the Department of General Zoology at the University of Duisburg-Essen with the idea of initiating a study on the reproductive biology of sengis after I have had the unique opportunity of observing short-eared sengis during my time as curator at Wuppertal Zoo. It occurred to me that it was time to start a serious data collection on this amazing little creature beyond of some anecdotal information which I had noted from time to time. Prof. Burda as a small mammal specialist has always been fascinated by these unusual animals and he agreed to supervise this project. I thank him for all the advice, friendship and interest that I received from him.

But first of all, my husband Dr. Alexander Sliwa deserves very special thanks for encouraging me to start this research project, providing active and moral and support and for having put up with me during my dissertation. Being an ambitious zoologist too, he inspired me many times with useful ideas in the course of numerous discussions, reviewed some early drafts and helped to design various illustrations. He followed my research progress with big interest and reported some valuable information.

At the same time I would like to thank my parents, Drs. Franz and Adelheid Olbricht, who always motivated me to fight my way through, gave emotional support and helped with the children.

Wuppertal Zoo provided me with generous access to my study animals, the sengis, and I would like to extend my sincerest thanks to Dr. Ulrich Schürer and his staff. Especially, the keepers of the Great Apes Department and the Felid Department were always there to advise me and provided me with valuable details of sengi husbandry and additional information on my study animals during my absence. From these departments I obtained a large number of deceased specimens. Here again my husband as the former curator at Wuppertal Zoo, Zoo veterinarian Dr. Arne Lawrenz and curator André Stadler were excellent supporters in terms of communication and technical support. I had the invaluable opportunity to share the zoo's data bank for sengis since 1976.

I am most grateful to Beryl Wilson, collection manager of the McGregor Museum in Kimberley, South Africa who shared her experience on sengis in the field with me and the data on various sengi species in the museum collection. Together with Lucas Thibedi of the Amathole

Museum in King William's Town, South Africa, she measured some specimens available in their collections. I am also grateful to Dr. Gustav Peters of the Zoological Institute and Museum Alexander Koenig, Bonn, Germany, Wim Wendelen of the Department African Zoology of the Royal Museum for Central Africa in Tervuren, Belgium and Dr. Siegfried Czernay and Jutta Heuer of Halle Zoo, Mr. Andreas Filz of Tierpark Bernburg, and Roy Bäthe from Erfurt Zoo, all Germany, for access to their collections or donations of specimens. The Natural History Museum of the Humboldt University Berlin greatly contributed by donating a female *Petromus typicus* which was crucial to define the position of dorsal teats. I thank Christine Bartos and the staff from the sengi department at Philadelphia Zoo for the good communication regarding data for *Rh. petersi*. I gratefully acknowledge the skillful help in preparing and staining the histological sections by Astrid Sulz from the Institute for Anatomy, University of Munich. Rebecca Banasiak from the Field Museum of Natural History in Chicago, USA, provided the schematics of Figs. 2.1 and 3.1 and constructed Fig. 3.2 based on photographs.

I am greatly indebted to the following scientists and the co-authors of publications which will derive from the data of this thesis. Prof. Mike Perrin of the University of KwaZulu-Natal, Pietermaritzburg, RSA, Dr. William Stanley from the Field Museum of Natural History in Chicago, USA, Prof. Ulrich Welsch from the Institute for Anatomy, University of Munich, Dr. Robert Asher from the Museum of Zoology at the Cambridge University, UK, who provided additional information and advice for various aspects of the manuscript. With their scientific experience and enthusiasm they were highly supportive. Also special thanks to Dr. Galen Rathbun of the California Academy of Sciences, San Francisco, USA, and to Klaus Rudloff, curator at Tierpark Berlin-Friedrichsfelde, for the inexhaustible dialogue on the Afrotheria in general and sengis in particular.

So many others have helped with the work or just helped to keep me sane. To all the graduate students in the Department of General Zoology of the University Essen-Duisburg: thanks for all the good times and all the help they have given me over the years. In particular, Dr. Sabine Begall was always there to be of assistance with statistic and special questions or just to relax my mind in times of frustration. Together with Dr. Regina Moritz she was also a never dwindling source for providing scientific articles and good spirits

LIST OF CONTENTS

ACKNOWLEDGEMENTS

I   **SUMMARY** — 1

II   **ZUSAMMENFASSUNG** — 3

III   **GENERAL INTRODUCTION** — 5

III.1   Fossil and extant macroscelids — 5
III.2   The Afrotheria - hypothesis in mammalian evolution? — 6
III.3   Testicond afrotheres — 7
III.4   Molecular versus morphological research — 7

CHAPTER 1

1. REPRODUCTIVE PATTERNS OF SENGIS (MACROSCELIDEA) – A THEORETICAL APPROACH — 8

1.1   Abstract — 8
1.2   Introduction — 8
1.2.1   Aim of the study — 9
1.3   Material and Methods — 11
1.4   Results — 11
1.5   Discussion — 14
1.5.1   Placentation in afrotheres — 15
1.5.2   Developmental status at birth — 15
1.5.2.1   Environmental factors influencing the developmental status of birth —16
1.5.2.2   Development of precociality — 16
1.5.2.3   Precocial and altricial strategies — 17
1.5.3   Gestation length — 18
1.5.4   Parental care — 18
1.5.4.1   Lactation — 19
1.5.4.2   Feeding intervals — 19
1.5.5   Monogamy — 19
1.5.5.1   Monogamy and parental care — 21
1.5.5.2   Mate guarding — 20

1.5.5.3	Monogamy and precociality — 21
1.5.6	Juvenile mortality — 21
1.5.7	Litter size — 21
1.5.7.1	Correlation between teat number and litter size — 22
1.5.7.2	Limits of litter size — 23
1.5.8	Seasonality of breeding periods — 24
1.5.8.1	Male capacities — 25
1.5.9	Post-partum oestrous — 25
1.5.10	Poly-ovulation — 25
1.5.10.1	Female capacities — 26
1.5.11	Longevity and fecundity — 26
1.6	Conclusions — 27

CHAPTER 2

2. THE TOPOGRAPHIC POSITION OF THE PENIS IN SENGIS (MACROSCELIDEA); AND COMMENTS ON PENIS TOPOLOGY IN TESTICOND MAMMALS — 30

2.1	Abstract — 30
2.2	Introduction — 30
2.2.1	Sengis as testicond afrotheres — 31
2.1.2	Taxonomic tools — 31
2.1.3	Aim of the study — 32
2.2	Material and Methods — 32
2.2.1	Material — 32
2.2.2	Methods — 32
2.2.2.1	Hyracoidea — 33
2.2.2.2	Statistical methods — 34
2.3	Results — 34
2.3.1	Sengi measurements — 34
2.3.2	Hyrax accounts — 37
2.4	Discussion — 37
2.4.1	Distance anus-penis — 37
2.4.2	Genital morphology in the Afrotheria — 38
2.4.3	Copulation posture — 38
2.4.4	Hyracoidea — 39
2.5	Conclusions — 39

CHAPTER 3

3. THE TAXONOMIC DISTRIBUTION OF MAMMARY GLANDS IN SENGIS — 40

3.1 Abstract — 40
3.2 Introduction — 40
3.2.1 Nomenclature of teat positions — 41
3.2.2 Behavioural ecology — 42
3.2.3 Aim of the study — 42
3.3 Material and Methods — 42
3.3.1 Material — 42
3.3.2 Methods — 42
3.4 Results — 44
3.4.1 Locations of the mammae — 44
3.4.1.1 Males — 44
3.4.1.2 Females — 45
3.4.1.3 Formulas — 49
3.5 Discussion — 50
3.5.1 Teats on male sengis — 50
3.5.2 Teats on female sengis and formulas — 51
3.5.3 The enigma about nuchal, lateral and dorsal teats — 52
3.5.3.1 Historical observations — 52
3.5.3.2 The presence of dorsolateral teats — 52
3.5.3.3 Defining a dorsolateral teat location — 53
3.5.3.4 Functionality of teats — 53
3.6 Conclusions — 54

CHAPTER 4

4. HISTOLOGICAL AND HISTOCHEMICAL STUDY OF THE MAMMARY GLANDS OF *PETRODROMUS TETRADACTYLUS* — 55

4.1 Abstract — 55
4.2 Introduction — 55
4.2.1 Mammae in male mammals — 56
4.2.2 Aim of the study — 56
4.3 Material and Methods — 57

4.3.1 Material — 57
4.3.2 Methods — 57
4.3.2.1 Fixation, embedding and staining methods — 57
4.3.2.2 Histochemistry of lectins — 58
4.4 Results — 58
4.4.1 General findings (light microscopy) — 58
4.4.1.1 Females — 58
4.4.1.2 Males — 58
4.4.2 Specific histological and histochemical findings — 59
4.4.2.1 Female mammary gland — 59
Actin — 59
PAS reaction — 59
Alcian blue — 59
Lectins — 61
4.4.2.2 Male mammary gland — 63
Lectins — 63
4.4.2.3 Scent glands — 65
4.5 Discussion — 67
4.5.1 Histo-morphology of mammary glands in *Petrodromus* — 67
4.5.2 The presence of mammae in male sengis — 67
4.5.3 Other glands — 68
4.6. Conclusions — 69

CHAPTER 5

5. BODY METRICS OF THE SHORT-EARED SENGI (*MACROSCELIDES PROBOSCIDEUS*, SMITH 1829) — 70

5.1 Abstract — 70
5.2. Introduction — 70
5.2.1 Body metrics and other physical patterns — 71
5.2.2 Growth models — 71
5.2.3 Aim of the study — 72
5.3 Material and Methods — 72
5.3.1 Material — 72
5.3.2 Methods — 72
5.3.3 Statistical analysis — 73

5.3.3.1		The Gompertz growth model — 74
5.4	Results — 74	
5.4.1		The Gompertz model — 74
5.4.2		Sexual dimorphism — 76
5.4.3		Correlation of body length with hind foot length and body mass — 77
5.5	Discussion — 78	
5.5.1		Measuring scheme and statistical methods — 78
5.5.2		Sexual dimorphism — 79
5.5.3		Developmental stage at birth and mating system — 79
5.5.4		Growth patterns and maturity — 80
5.5.4.1		The Gompertz growth parameters — 80
5.5.4.2		Body length against hind foot length and body mass — 80
5.5.4.3		Factors influencing precociality — 81
5.5.4.4		Lactation period — 82
5.5.4.5		Skeletal growth and sexual maturity — 83
5.6	Conclusions — 85	

CHAPTER 6

6. POST-NATAL BODY MASS DEVELOPMENT OF THE SHORT-EARED SENGI (*MACROSCELIDES PROBOSCIDEUS*) — 84

6.1	Abstract — 84	
6.2	Introduction — 84	
6.2.1		Growth models — 84
6.2.2		Aim of the study — 85
6.3	Material and Methods — 85	
6.3.1		Material — 85
6.3.2		Methods — 85
6.3.2.1		Gompertz growth model and statistics — 86
6.4	Results — 86	
6.4.1		Growth curves — 86
6.4.2		Gompertz growth parameters — 87
6.4.2.1		Sex-specific growth — 88
6.5	Discussion — 89	
6.5.1		Environmental and behavioural impact on weight development — 89

6.5.2		Growth parameters — 90	
6.5.3		Adult body mass — 91	
6.5.3.1		Estimated adult weight — 91	
6.5.3.2		Relating body mass to sexual maturity — 92	
6.5.3.3		Sexual dimorphism — 92	
6.6	Conclusions — 92		
IV	**REFERENCES**	**—**	**93**
V	**APPENDIX**	**—**	**107**
A	List of tables — 107		
B	List of figures — 108		
C	Staining procedures — 109		
D	Table: Raw data on individual body measurements of male and female *Macroscelides proboscideus* — 112		

I SUMMARY

Sengis have been studied for more than a century but information on their biology is still scattered. The reproductive biology of sengis is best understood in the context of their evolutionary history. Their phylogeny has long been the subject of much speculation and controversy. This thesis aims to consolidate molecular findings of other studies with morphological methods and thus, to contribute to a better understanding of their phylogeny. Sengis are members of the Afrotheria, an African clade of mammals, and particular attention was paid in this study to their relationships to other afrotheres.

Chapter 1 summarizes the knowledge on reproductive parameters across the order Macroscelidea from the literature. Various reproductive characteristics are unusual for a small mammal. The most important traits in the life history of sengis in terms of reproduction are precociality and monogamy as well as the ability of some species to poly-ovulate and to perform post-partum oestrus. Further, they have a very long lifespan with high fecundity rates.

In chapters 2 and 3 I apply morphological methods to investigate phylogenetic relationships. Morphological landmarks such as the position of penis and teats were defined. The position of the penis is of utility in distinguishing between the genera *Petrodromus* and *Macroscelides*, but not between any other genera of sengi, supporting recent taxonomic conclusions regarding the relationship of these two taxa. Teat position is useful in taxonomic distinction among the three of the four genera, only between *Macroscelides* and *Elephantulus* there was no difference. Only in two species teats were found on males: *Elephantulus rozeti* and *Petrodromus tetradactylus* but not on other species examined. This supports the recent taxonomic conclusions regarding the relationship of these two taxa. The arrangement of teats was determined in a formula for each genus which distinguishes between antebrachial, abdominal and inguinal regions. No sengi exhibited teats situated dorsolaterally to the extent of those in rock-dwelling mammals such as *Petromus typicus*.

In chapter 4 mammary tissue of a female and male *P. tetradactylus* was examined with different histological and histochemical methods. The results reveal a potentially functioning mammary gland in male *Petrodromus* with evidence of active mammary tissue. The secretory units (acini) are sexually dimorphic. In the female typical acini, milk ducts, cisternal milk sinus and a teat canal can be distinguished. The acini of the females occur in the periphery of the gland whereas acini in the male teat occur in the connective tissue of the teat. The function of mammary tissue in male *Petrodromus* is not clear because males of none of the sengi species contribute to the raising of their young. Apocrine scent glands were found in both genders at the base of the teat which underlines the importance of chemical communication for sengis.

SUMMARY

In chapter 5 body measures (mass, length of body, head, ear, snout, whiskers, tail, hind foot) of captive short-eared sengis (*Macroscelides proboscideus*) were taken post mortem and then fitted to the 3-parameter Gompertz model There was considerable variation of the growth parameters of these body measures in terms of growth constant (K) and inflection age (I). Whiskers and snout had the fastest growth, the ears the slowest. The asymptotic value of the growth model (A) in terms of adult length of tail and ear as well as body mass was exceeded later then the sigmoidal curves suggested but nevertheless, adult size of all body parts is achieved at about sexual maturity (ca. 45 days), except hind foot length which reached its maximum earlier. No significant sexual dimorphism in the estimated adult size could be determined.

Chapter 6 refines the results regarding body mass growth during the ontogeny of individual short-eared sengis which were weighed on a nearly daily basis from their first days of life until adulthood. The Gompertz growth model was used to generate the growth parameters K, I and A which were compared with data on the reproductive biology of sengis. Furthermore, the growth parameters for *Macroscelides* were compared with those of various species obtained from the literature. Adulthood is reached when adult size matches with sexual maturity, at about 45 days. There were no significant differences between males and females in growth or adult body mass size.

II ZUSAMMENFASSUNG

Rüsselspringer oder Elefantenspitzmäuse (Macroscelidea) werden seit mehr als einem Jahrhundert wissenschaftlich untersucht, jedoch sind Informationen zu ihrer Biologie lückenhaft und zusammenhanglos. Die Fortpflanzungsbiologie von Rüsselspringern wird am besten im Kontext zu ihrer Evolutionsgeschichte verständlich. Ihre Phylogenie war lange Gegenstand von Spekulationen und Kontroversen. Diese Arbeit hat zum Ziel, Erkenntnisse aus der Molekulargenetik mit morphologischen Methoden zu unterstützen und somit zu einem besseren Verständnis ihrer Stammesgeschichte beitragen. Rüsselspringer gehören zu den Afrotheria, einer Gruppe von endemischen afrikanischen Säugetieren. Diese Verwandschaftsbeziehungen fanden hier besondere Berücksichtigung.

Kapitel 1 faßt das bisherige Wissen über Aspekte der Fortpflanzungsbiologie der Ordnung Macroscelidea zusammen. Einige Charakteristika sind für Kleinsäuger ungewöhnlich. Zu wichtigen Reproduktionsparametern zählt, dass Rüsselspringer Nestflüchter sind und monogam leben. Daneben sind einige Arten zu Polyovulation und post-partum- Östrus befähigt. Während ihrer relativ langen Lebensdauer erreichen sie eine bemerkenswerte Geburtenrate.

In den Kapiteln 2 und 3 wandte ich morphologische Methoden zur Untersuchung phylogenetischer Verwandschaftsbeziehungen an. Morphologische Meßpunkte wie die Position von Penis und Zitzen wurden festgelegt. Die Position des Penis diente der Unterscheidung der Gattungen *Petrodromus* und *Macroscelides*, jedoch nicht zu oder zwischen den anderen Gattungen. Dieses Ergebnis bestätigt neuere taxonomische Schlussfolgerungen zur weniger engen Verwandtschaft dieser beiden Taxa. Die Lage der Zitzen führte zur taxonomischen Unterscheidung von 3 der 4 Gattungen, zwischen *Elephantulus* und *Macroscelides* besteht kein bedeutsamer Unterschied. Nur bei den beiden Arten *Petrodromus tetradactylus* und *Elephantulus rozeti* wurden beim Männchen Zitzen gefunden, was Ergebnisse neuerer Untersuchungen zur engen Verwandtschaft dieser beiden Taxa bestätigte. Die Anordnung der Zitzen definierte ich für alle 4 Gattungen in einer gattungsspezifischen Formel, wobei zwischen antebrachialer, abdominaler und inguinaler Position unterschieden wurde. Kein Rüsselspringer zeigte dorsolaterale Zitzen, wie sie bei der Felsenratte (*Petromus typicus*) bekannt sind.

Das Milchdrüsengewebe einer männlichen und weiblichen Rüsselratte (*P. tetradactylus*) wurde mit verschiedenen histologischen und histo-chemischen Methoden in Kapitel 4 untersucht. Die Egebnisse ergaben beim Männchen eine funktionsfähige Milchdrüse mit aktivem Milchgewebe. Die Drüsenendstücke unterschieden sich geschlechtsspezifisch. Beim Weibchen fanden sich typische Drüsenendstücke, Milchgänge sowie Milchsinus und Zitzenkanal. Die Drüsenendstücke des Weibchens befanden sich in der Drüsenperipherie, die des Männchens im Bindegewebe der

Zitze. Die Funktion von Milchdrüsengewebe bei der männlichen Felsenratte bleibt unklar, da sich bei keiner Rüsselspringerart die Männchen an der Jungenaufzucht beteiligen. Apokrine Duftdrüsen wurden bei beiden Geschlechtern an der Zitzenbasis gefunden, was die große Bedeutung der olfaktorischen Kommunikation für Rüsselspringer unterstreicht.

Die Kapitel 5 und 6 beschäftigen sich mit post-natalem Wachstum. In Kapitel 5 wurden Körpermaße wie die Längen von Kopf-Rumpf, Ohr, Schnauze, Fibrissen, Schwanz und Hinterfuß sowie das Körpergewicht post-mortem bei zoogeborenen Kurzohr-Rüsselspringern (*Macroscelides proboscideus*) verschiedenen Alters vermessen und mit Hilfe des Gompertz-Wachstumsmodells analysiert. Dabei konnten erhebliche Unterschiede beim Wachstum der verschiedenen Körpermaße im Hinblick auf die Wachstumskonstante (K) und den Zeitpunkt des schnellsten Wachstums (I) festgestellt werden. Bei Fibrissen und Schnauze konnte generell das schnellste Wachstum verzeichnet werden, beim Ohr das langsamste. Die Annäherung an die Asymptote (A) für die adulte Länge von Schwanz und Ohr, sowie für das Adultgewicht zeigte sich in Anwendung des Wachstumsmodells später als es der sigmoidale Kurvenverlauf der direkten Meßwerte vermuten ließ. Der Hinterfuß erreichte seine Adultlänge früher als die anderen Körperteile, für die Adultmaße ungefähr beim Einsetzen der Geschlechsstreife (ca. 45. Lebenstag) ermittelt wurden. Keine Signifikanz konnte für geschlechtsspezifischen Dimorphismus der im Modell ermittelten geschätzten Adultmaße festgestellt werden.

Die tägliche individuelle Gewichtszunahme von zoolebenden Kurzohr-Rüsselspringern von den ersten Lebenstagen an bis ins Erwachsenenalter war Gegenstand der Untersuchung in Kapitel 6. Die Ergebnisse des vorherigen Kapitels konnten auf der Grundlage von kompletten Meßreihen der Gewichtsentwicklung überprüft und mit Hilfe des Gompertz-Modells analysiert werden. Die Wachstumsparameter K, I und A wurden mit Aspekten der Fortpflanzungsbiologie von Rüsselspringern in Zusammenhang gesetzt und konnten gleichzeitig mit den für andere Arten aus der Literatur bekannten gleichen Wachstumsparametern verglichen werden. Es gab keine signifikanten Unterschiede zwischen Männchen und Weibchen bezüglich Wachstum oder Adultmaß beim Körpergewicht. Das Ergebnis für das Erreichen des Adultgewichtes entsprach mit 45 Tagen dem Ergebnis des vorherigen Kapitels.

III   GENERAL INTRODUCTION

The Macroscelidea is a monophyletic order with the family Macroscelidae which comprises 16 well defined species in two subfamilies (Rovero et al. 2008): the Macroscelidinae with the genera *Petrodromus*, *Elephantulus* and *Macroscelides* and the Rhynchocyoninae with the only genus *Rhynchocyon*.

The English name "elephant-shrew" originated because of the superficial similarities the Macroscelidea bear to other mammals, and in historical (and erroneous) taxonomic opinions. In particular, these animals have a very different evolutionary history than true shrews (Soricidae) and they share few life history traits. Because of this misleading nature of this name, authors (e.g. Rathbun and Woodall 2002, Skinner and Chimimba 2005) are increasingly using the African term "sengi" for elephant-shrew, which is derived from the Kiswahili word "sanje" (Eastern Africa, Rathbun and Kingdon 2006) or from the Lunda-word "Isengi" (Zambia, White and Ansell 1966). I agree with these scientists that it is appropriate to use local names and this protocol is followed here.

III.1   Fossil and extant macroscelids

The macroscelidean fossils from the Eocene and Oligocene (55-23 Ma) showed the early diversity in this group and thus provided new arguments for the origin and interordinal relationships of the Macroscelidea (Simons et al. 1991). Possibly, already during the early Eocene (33 Ma), ancestors of macroscelids branched off. The divergence of the two subfamilies may date from earlier Oligocene (Patterson 1965). The Rhynchocyoninae have undergone only minor evolutionary change since the early Miocene (Novacek 1984) at which time the ancestors of the Macroscelidinae were already highly specialized. The extant Rhynchocyoninae are represented by the single genus *Rhynchocyon* which stands apart from the Macroscelidinae in a number of characters (Corbet and Hanks 1968). It occurs in eastern and central Africa whereas the three macroscelidine genera are widely distributed from North Africa (only *Elephantulus rozeti*) to eastern and southern Africa.

The extinct forms of sengis were herbivorous (Patterson 1965) which is considered a plesiomorphic character and a link to their paengulate relatives (Hyracoidea, Proboscidea and Sirenia). Insectivory is a later adaptation (Kerley 1995). Sengis are omnivorous, although believed to be exclusively insectivorous for a long time (Kerley 1989, 1995) and for a long time the family Macroscelidae had been placed within the former order Insectivora (Starck 1948, Nowak and

Paradiso 1983). Butler (1956) revised the Insectivora after comparisons with the fossil skull of Ictops and changed the status of the former family Macroscelididae into the new order Macroscelidea. Later this group was associated with the Tupaiidae (Haeckel 1866) and aligned with lagomorphs and rodents (McKenna 1975, Coldiron 1977, Novacek 1992) but no definite judgement as to possible relationships could be made with the results of these studies.

A biogeographic review of sengis and the need for their conservation were published in the Databank for the Conservation and Mangement of the African Mammals (Boitani et al. 1999) and the IUCN Action Plan (Nicoll and Rathbun 1990) where for six species or subspecies intermediate levels of threat were estimated.

III.2 Afrotheria – hypothesis in mammalian evolution?

The Afrotheria, endemic Afro-Arabian group of placental mammals includes among sengis morphologically very diverse forms like golden moles (Chrysochloridae), tenrecs (Tenrecidae), aardvarks (Tubulidentata), hyraxes (Hyracoidea), elephants (Proboscidea), and the dugongs or manatees (Sirenia) (Robinson and Seiffert 2004). The Proboscidea, Hyracoidea and Sirenia are known as Paengulata (Novacek 1992), the Chryochloridae and Tenrecidae have been also grouped together as Afrosoricida (Robinson and Seiffert 2004) or Tenrecomorpha (Carter et al. 2004). De Jong et al. (1981) were the first who suggested that aardvarks are closely related to sengis, elephants, sea cows and hyraxes based on the study of alpha A-crystallin. Later, golden moles (Stanhope et al. 1998a) and tenrecs (Stanhope et al. 1998b) were included and consequently, the entire group was given the name Afrotheria.

The Cretaceous afrotherian ancestor likely was a small forest-dwelling insectivore or possibly herbivore (Hedges 2001). During the mid-Cretaceous period (105-90 Ma) Africa was isolated and the diversification of the mammals - now included in the superordinal clade Afrotheria - proceeded (Stanhope et al. 1998b). The crown (extant) afrotherians appeared approximately 80 Million years ago after the stem afrotherians had about 25 million years of independent evolution (Springer et al. 2003). It has been suggested that sengis represent the earliest African lineage (Stanhope et al. 1998a).

Despite of the strong molecular support for afrotherian monophyly (de Jong et al. 1981, Stanhope et al. 1996, 1998a, b; Springer et al. 2003; van Dijk et al. 2001; Nikaido et al. 2003) morphological features shared by these animals, particulary between the insectivoran- and ungulate-grade afotherians, have been difficult to identify (Asher and Lehmann 2008). Several similarities became apparent only recently, including non-descent of the male gonads (Werdelin and Nisonne 1992), morphology of the placenta (Mess and Carter 2006), variable vertebral count (Sánchez-

Villagra et al. 2007), a concave cotylar facet of the astragalus (Tabuce et al. 2007), calcano-navicular contact (Seiffert 2007) and dental eruption (Asher and Lehmann 2008). Nishihara et al. (2006) concluded that the afrotherian lineages diverged very rapidly with the consequence that ancestral polymorphism present in the last common ancestor of Paenungulata results in incongruence. The diversification of the Macroscelidea included both slow and accelerated morphological evolution (Douady et al. 2002).

One of the most pressing interpretive problems is still the uncertain phylogenetic position of the two orders aardvarks and sengis within the Afrotheria (Seiffert 2002). Sengis preserve a number of specialized paenungulate-like features in their postcranial, cranial and dental morphology but various genetic studies support close relationship of sengis with aardvarks, tenrecs and golden moles (Waddell et al. 2001).

III.3 Testicond afrotheres

Seiffert (2002) suggested that Afrotherians evolved from a common ancestor with a very primitive (or secondarily primitive) reproductive system, most male afrotheres are more similar to monotremes than to marsupials and most other placentals in being primary testicond. Testicondy, the retention of testes in their ontogentically primary position in the abdominal cavity is a derived condition among therians and has been viewed as afrotherian synapomorphy which provides morphological support for afrotherian monophyly (Werdelin and Nilsonne 1999).

III.4 Molecular versus morphological research

Sengis have attracted attention of researchers for over more than a century. In recent years, systematists have been struggling to reconcile classical "morphological" methods of reconstructing evolutionary trees based on anatomical similarities and differences between living species or their extinct relatives with an avalanche of new molecular data on genetic variation among organisms (Gibbons 1991, Asher et al. 2003). As a result, the face-off between proponents of molecules and those of morphology was sometimes controversial. Balter (1997) stresses the necessity that systematists of both camps must move closely together to sort out the many remaining puzzles since only a few groups of animals have had their phylogenetic trees worked out with complete confidence and that there are warning signs that molecular evidence can lead to misleading errors. In contrast Hedges (cited as pers. comm. in Balter 1997) maintained that morphological features are more susceptible to "adaptive convergence" (homoplasy) and consequently, trees should be built with molecular data alone. This research supports molecular findings with morphological methods and is consolidating the knowledge on phylogenetic relationships.

CHAPTER 1

1. REPRODUCTIVE PATTERNS OF SENGIS (MACROSCELIDEA) – A THEORETICAL APPROACH

1.1 Abstract

This theoretical approach summarizes the knowledge on reproductive parameters across the order Macroscelidea. Morphological parameters such as teat number were related to life history traits such as litter size, gestation length, developmental status at birth and mating system. The most important traits in the life history of sengis in terms of reproduction are precociality and monogamy as well as the ability of some species to poly-ovulate and to perform post-partum oestrus. Sengis have a very long lifespan. Particular attention was paid to similarities among afrotherian mammals.

1.2 Introduction

Nicoll and Rathbun (1990) emphasized the need for more research since sengis are not well-known biologically and more knowledge is required for better understanding and more effective conservation. Valuable information on sengis in the wild as well as in captivity is available but there is a need to pool and summarize this knowledge to connect different parameters and subsequently to detect interrelations. Relationships between reproductive traits and life history in the Macroscelidea have not been reviewed yet. Most of the characters here are compared with other mammals such as rodents due to the lack of information on sengis and if available with other afrotheres. Tabuce et al. (2008) stress the necessity of the development of new sources of phylogenetic characters to shed light on the Afrotheria.

Traits such as mammary numbers may be an important factor in the evolution of litter size (Pearl 1913a; Gilbert 1986). The correlation between mean numbers of mammae and mean litter sizes in rodents is described as the "one-half rule" (Gilbert 1986, Sherman et al. 1999). The mean litter size was about one half of the mean number of mammae whereas the maximum litter sizes were approximately equal to the number of mammae. Summarizing his results, Gilbert (1986) came to the conclusion that mammary number constrains litter size rather than vice versa.

Considering the fact that most macroscelids have precocial young (e.g. Skinner and Smithers 1990) my interest is devoted to understanding the adaptive significance of their reproductive patterns. In addition, the energy and nutrient requirements of reproduction will be influenced by such characteristics as litter size, stage of development at birth and the duration of gestation and lactation (Neal 1986).

In particular in the Macroscelidinae it is well documented that they have a relatively large number of teats in relation to small litters. This central issue is followed throughout the entire investigation and required a closer look on causal directions of correlations playing a role in this relationship. The question arose whether morphological traits influence lifestyle or whether lifestyle requires changes of these traits.

Generally the discussion of causal directions of the correlates is difficult. At this point Weir and Rowlands (1973) in their review on reproductive strategies of mammals can be cited as follows: "the complexity of known mammalian reproduction is such as to confuse what may once have been the original pattern, if indeed there ever was just one. Many features in this respect do coincide, rather than there are differences from one species to the next generic, familial, or ordinal relationships".

1.2.1 Aim of the study

The objective of this investigation was to find cross-species correlations between morphological parameters such as mammary number, life history traits such as litter size, developmental status of neonates (precocity or altricity), gestation length or maximum lifespan with regard to different lifestyles. Particular attention was paid to discuss reproductive traits of sengis in the context of connections and continuity within the Afrotheria and to provide a critical appraisal of the macroscelids´ position within this superorder.

The following hypotheses were formulated regarding the relatively large number of teats in relation to small litter size:

1. it can be explained by phylogenetic inertia;

2. it can be related to the absent maternal care system and possibly to social monogamy.

Fig. 1.1. The four sengi genera: (top) *Rhynchocyon petersi*, Denver Zoo (photo: A. Sliwa); (center left) *Macroscelides proboscideus*, Wuppertal Zoo (photo: A.Sliwa) (center right) *Elephantulus brachyrhynchus*, Transvaal Snake Park (photo: A.Sliwa); (bottom) *Petrodromus tetradactylus*, Museum of Zoology, Cambridge (photo: G. Olbricht).

1.3 Material and Methods

An extensive bibliography is available on the Macroscelidea and many research fields are well documented. The wide range of sources over more than a century proved to be very suitable for conducting a review. Early works sometimes are controversial compared to more recent research, which allows a critical appraisal. Data on all the parameters used for Table 1.1 were drawn from the extensive published literature, Fig. 1.1 presents members of all four sengi genera.

1.4 Results

For 14 of the 17 extant sengi species of Macroscelidea (Rathbun 2009) nine life history or reproductive traits are listed in Table 1 which could be of importance for a critical appraisal of the hypotheses.

No data were available for *E. fusciceps, E.pilicaudus* and *R. udzungwensis*. Differences between the genera become evident in body mass, ovulation type, gestation time and, number of offspring.

One of the diagnostic characters distinguishing the two subfamilies is the shape of the uterus. *Rhynchocyon* has a slightly bicornate uterus whereas its shape in all the other sengis is deeply bicornate (Corbet and Hanks 1968). Morphological differences also include teat number; while the Macroscelidinae except *Petrodromus* have 6 mammae, the Rhynchocyoninae and *Petrodromus* have 4. More details and modifications are presented in Chapter 3 which is dedicated to teat location. The special ultrastructure of the spermatozoa (Woodall and FitzGibbon 1995) is considered an ancestral feature, which divides between sengi genera.

Most sengi species breed year-round but some show seasonal trends with reduced reproduction in the cool season, such as *E. brachyrhynchus* (Neal 1995). Singletons or twins are standard with the exception of *E. rozeti* where three and four young are common in some areas (Séguignes 1989). In *Macroscelides* some exceptional birth of triplets were noted in captivity by Olbricht et al. (2006). All members of this order produce precocial young except for *Rhynchocyon* which bear less developed offspring. Eye opening in *R. petersi* is at 11 – 12 days old (Lengel 2007) while the other sengis are born with open eyes. All sengis are considered monogamous (Brown 1964, Rathbun 1979a, Kingdon 1984; Woodall and Skinner 1989, Ribble and Perrin 2005). They live in pairs or singly and seldom in small groups.

Table 1.1 Life history characters of the Macroscelidea (pect. = pectoral, abd. = abdominal).

Species	Mass of adult (of neonate) in g	Gestation in days	No. of young (of fetuses)	No. of mammae	Ovulation type	Breeding season *	Longevity in years
Rhynchocyon chrysopygus (Golden-rumped sengi)	540 (80)[1]	42[1]	1[1]	4 (abd.**)[8]		year-round[1]	11[20]
R. cirnei (Chequered sengi)	420[15]		(2)[15]	4 (abd.)[8]		year-round[18]	
R. petersi (Black-and-Rufous sengi)	max.740[2] (34)[2]	45-47[2]	(2)[13] 2[2]	4 (abd.)[8]	oligo[13]		4,5[20]
Petrodromus tetradactylus (Four-toed sengi)	232 (31)[9]		max.2[5,18] (1)[13]	4 (pect., abd.)[8] (absence of abd. pair,[14])	oligo[13]	Dec[3], June[21], Jan + July[19], whole year[18]	6.5[20]
Macrsocelides proboscideus (Short-eared sengi)	ca. 40[5] 46[22](ca.8)	60-61[4]	max3[4]	6 (pect, abd. inguinal)[3,6,8,14]	poly[13]	year-round but trends (Sep-Feb)[12]	8[20]
Elephantulus brachyrhynchus (Short-snouted sengi)	52[5] (max 9)[27]		max 2[3], (1-2)[13]	6[14]	intermediate[13]	year-round[10], June, Nov[3], Oct-March[5,]	4[20]
E. edwardii (Cape sengi)	52[5], 48[6] (9)[6], (11,9)[17]		max 2[5,6] (2)[13]	6[14]	poly[13]	Nov-Jan[5], Sep-Dec[6,28]	5.5[20]
E. fuscus (Dusky e-shrew)	no data	no data	no data	no data	no data	probably like E. brach.[5]	no data
E. intufi (Bushveld sengi)	49[9],56[5] (10)[9]	50-52[9]	2[3], (1-2)[13]	6[3,14]	intermediate[13]	Nov, March[3], Aug-March[5]	9[20]
E. myurus (Eastern rock sengi)	69[9], 80[5] (8)[9]	61 interbirth interv.[23]	(2)[13]	6[14]	poly[13]	July-Jan[11], Sep-March[5]	
E. revoili (Somali sengi)				6[16]			
E. rozeti (North African sengi)	41[9]	75[7]	1-4[7] (1-4)[13]	6[14]	oligo[13]	Jan-Aug[13]	7[4]
E. rufescens (Rufous sengi)	max 55[24],(10,6)[26]	57[26] 61-65[28]	(1-2)[13]		oligo[13]	year-round[24], seasonal rains[21]	almost 8[20]
E. rupestris (Western rock sengi)	max. 68[5]		1-2[3,24]	6[3,14]	oligo[13]	year-round[25], Sep,April, May,[5]	4[20]

References in Table 1.1
1 Rathbun 1979a
2 Lengel 2007
3 Shortridge 1934
4 Olbricht et al. 2006
5 Skinner and Smithers 1990
6 Stuart et al. 2003
7 Séguignes 1989
8 Fitzinger 1867
9 Tripp 1972
10 Leirs et al. 1995
11 van der Horst 1946
12 Bernard et al. 1996
13 Tripp 1971
14 Corbet and Hanks 1968
15 Ansell 1973

16 Haltenorth and Diller 1977
17 Dempster et al. 1992
18 Kingdon 1984
19 Nowak (Walker) 1991
20 Weigl 2005
21 Brown 1964
22 Rosenthal 1975
23 van der Horst 1954
24 Neal 1982
25 Withers 1983
26 Rathbun et al. 1981
27 Yarnell and Scott 2006
28 Koontz and Roeper 1983

Post-partum oestrous is common but direct observations revealed an oestrous cycle of about 2 or 3 weeks for free-ranging (Sauer and Sauer 1971) and 2 weeks for captive *Macroscelides* (Vakhrusheva 2000). Van der Horst (1946) investigated uteri and ovaries of *E. myurus jamesoni* and estimated the length from anoestrous to oestrous as about 3 weeks.

A modal estrous cycle of 13 days was estimated for *E. rufescens* with a range of 12 to 49 days (Lumpkin et al. 1982).

The longevity records are based on data from captivity and may represent unusually long-lived specimens. Rathbun (1979a) expected 4-5 years for *Rhynchocyon chrysopygus*, some years later a captive specimen reached the age of almost 11 years (Weigl 2005). This was the longest lifespan for a macroscelid ever recorded and is followed by *E. intufi* with 9 years and *M. proboscideus* and *E. rufescens* with about 8 years.

Patterns of behavioural ecology are closely related to reproductive patterns and, some of them are listed here. *Rhynchocyon* lives in coastal forests or in lowland and montane forests (Nicoll and Rathbun 1990). *Petrodromus* is likewise an inhabitant of coastal forests but also occurs in dense scrub and woodlands and in very dry scrub (Kingdon 1990). All the smaller sengis, the Macroscelidinae, prefer semi-arid savannah, bushland and woodland with *Macroscelides proboscideus* tolerating the most arid conditions of the Namib Desert (Sauer and Sauer 1971).

Ovulation occurs spontaneously. The Macroscelidae are divided into 3 ovulation groups although a gradation rather than a complete separation between species is more appropriate (Tripp 1971):

1. poly-ovulating species (31-89 corpora lutea = CL) include *Elephantulus myurus jamesoni*, *E. edwardii* and *M. proboscideus*,

2. intermediate form (1-23 CL): *E. intufi* and *E. brachyrhynchus* and

3. oligo-ovulating species (1-3 CL) are *Rhynchocyon petersi*, *Petrodromus*, *E. fusciceps*, *E. rupestris* and *E. rozeti*.

Surprisingly, the species with the most offspring among all the Macroscelidea, *E. rozeti*, is oligo-ovulating. In disagreement to Tripp (1971) who found in gravid *E. intufi* up to 8 corpora lutea (CL) and in *E. (Nasilio) brachyrhynchus* up to 23 CL per ovary, van der Horst (1944) found only one or 2 CL in both species. For *Petrodromus* and *E. rupestris* (oligo-ovulating) as well as for *Macroscelides* and *E. edwardii* (in van der Horst as *E. capensis*, poly-ovulating) both authors are in accordance. Additionally, van der Horst found in about half of the gravid females in *E. intufi* asymmetries of the uterus in contrast to this rare phenomenon in *E. myurus jamesoni*. In his study *E. intufi* carried an embryo in only one of the 2 horns but often 2 CL were found in one ovary. In *E. rupestris* 3 CL were present in the 2 ovaries. The chemical constitution of the egg is a remarkable point of distinction between sengi species that belong to the same genus, e.g. the presence and absence of fat-globules in the egg (van der Horst 1944). Fat-globules are absent in poly-ovulating species such as *Macroscelides* and *E. myurus*, and present, but in low numbers in the poly-ovulating *E. edwardii* and clearly present in oligo-ovulating species such as *E. rupestris* and *E. intufi*. (Tripp 1971).

Van der Horst (1954) confined the breeding season of *E. myurus jamesoni* between July and January. During his investigation females often ovulated at the end of the season without being fertilized which results in a menstrual cycle, as at the end of the season also the males go into anoestrous (Stoch 1954). This species is one of the relatively few mammals that seem to menstruate regularly (van der Horst and Gillman 1941, 1942b). This uterine bleeding is of an unusual type and was considered by Patterson (1965) as a macroscelidid peculiarity.

1.5 Discussion

A certain homogeneity of the analyzed traits can be generally stated despite of some interspecific variance. This is in accordance with Corbet and Hanks (1968) who noted the remarkable similarity in external morphology and Rathbun (1979b) and Perrin (1995a) who observed similar life history traits. Both patterns have remained remarkable static over geological time (Rathbun and Rathbun 2006a).

Due to the small litters low fecundity is expected, so that a premium is placed on traits that would increase their birth rates (Rathbun 1979b). In the following discussion I focus on potential factors controlling mammary number and litter size.

There is some confusion on teat number and position. The absence of abdominal teats is described only for *Petrodromus* (Fitzinger 1867; Corbet and Hanks 1968). Teat number of *Petrodromus* was misstated with three teat pairs instead of two (Tolliver et al. 1989) and for *E. myurus* with two instead of three (McKerrow 1954). Corbet (1995) worked out a cladogram on the base of synapomorphies but here he surprisingly also included *Petrodromus* in the group with abdominal teats together with *Rhynchocyon* with two teat pairs in the abdominal area (Fitzinger 1867). Teat location was not described in detail nor were there exact terms to define the position. Chapter 3 will investigate this issue in more detail.

The intergeneric variance in mammae number stimulates the discussion whether an analysis of adaptations should be useful or whether selection in progress is to be detected. An organism may have an adaptation even if selection in progress is not operating on it now or selection in progress may be modifying an existing adaptation, creating a new adaptation (Grafen1988).

1.5.1 Placentation in afrotheres

Sengis, together with rock hyraxes, tenrecs and golden moles show a more recent development, the hemochorial condition elephants and aardvarks exhibit the ancient endotheliochoral placenta of the common ancestors of the Afrotheria (Starck 1949, Carter and Enders 2004). The formation of a discoid placenta in sengis and tenrecs is still a further development. The lobulated allantoic sac, as part of the fetal membranes is a feature is possibly shared by all afrotherians whereas the proliferation layer has not been described for any other mammalian orders except for sengis (Oduor-Okelo et al. 2004). Some features of the placenta such as the inner degenerative zone are present in most sengis but are absent in *R. chrysopygus* (Carter et al. 2004).

Mammalian fetal membranes and placentas differ greatly without there being any obvious relation to life style and they are an alternate source of phylogenetic information (Carter et al. 2004). The effects of selection pressures on the efficiency of placentation may stem from changes in nutritional demand, gestation length, number of embryos per pregnancy and maternal body constitution (Wildman et al. 2006). Comparison of the placental type with the brain advancement factor reveals that the haemochorial placental type includes altricial as well as precocial groups (Sacher and Staffeldt 1974) but the complex relationships do not support any simple relation between placental type and gestation schedule. The example of the guinea pig (Eisenberg 1981), being also a precocial small mammal like most sengis, shows that species with haemochorial placentas pass almost the entire maternal antibody complement from the maternal circulation to the foetal circulation and pass little or none of it in the milk. Consequently, this would lead to highly developed young.

The significance of the lobation of the allantois is unknown (Oduor-Okelo et al. 2004), but amniogenesis by cavitation instead of folding is recognized as character transformation on the common stem lineage of Afrotheria which helps to identify morphological characters that could be synapomorphic for this novel taxon (Mess and Carter 2006). In addition, Carter and Mess recognize the four-lobed allantoic sac and precociality as two other characer transformations on the stem lineage of Afrotheria.

1.5.2 Developmental status at birth

Developmental length in general is a conservative trait in mammals which is primarily determined by phylogenetic factors and body size (Burda 1989b). A general controversial discussion is held about the classification itself if neonates are altricial or precocial because in reality mammalian species occur along a continuum of developmental maturity at birth (Zeveloff and Boyce 1986). Case (1978b) considers a mammal species precocial if its young possess body hair, teeth, open eyes, open auditory meatus, the ability to move forward, the ability to right itself when lying on its back and formed claws. Mammals lacking all or all but one ar altricial, those lacking some are semi-precocial or semi-altricial. In all sengis we find heavy and highly precocial neonates, except for *Rhynchocyon* which is considered semi-precocial following Case's definition for precociality. The developmental stage at birth at which a young is born is a more plastic trait than the developmental length (*Spalacopus cyanus*, Begall et al. 1999) or a conservative rather than a plastic trait (*Cryptomys sp.*, Begall and Burda 1998).

In their analysis of the correlation between maternal investment and litter size Zeveloff and Boyce (1986) demonstrate that mammals with litter sizes < 2 tend to bear precocial young and those with more than 3 young are altricial. This agrees with small litter size and precociality in the Macroscelidinae, hyraxes and precocial rodents such as chinchillas. Nevertheless they conclude that this does not imply that litter size should be a criterion for classification of neonate development.

1.5.2.1 Environmental factors influencing the developmental status at birth
Growth reflects an adaptation of species for their environment (Case 1978b). Predator selection might favour an increase in neonate size where young are reared on open, exposed areas (Case 1978a). This may have largely influenced the development of precociality of sengis living in dry habitats which do not build nests. The precocial young hyraxes are also quite exposed to predators although they live in crevices in the rocks.

Only *Rhynchocyon* occurring in dense forest can afford less advanced neonates. They build nests. On the other hand, the precocial rodent *Otomys* also constructs well developed nests and here the occurrence of precociality cannot be explained satisfactorily (Neal 1986). The altricial (or semi-precocial) tenrecs inhabit various habitat types (Eisenberg and Gould 1970, Puschmann 2004), only for some species nest building is reported, whereas the semi-precocial aardvark young does not leave its burrow for the first 2 weeks after birth (Shoshani et al. 1988). The altricial golden moles live in burrows and so are well protected (Perrin and Fielden 1999).

1.5.2.2 Development of precociality
There is some controversy on about the development of precociality. Following Hopson's theory (1973) short gestation periods of small placentals followed by altricial young and intensive maternal care, including suckling is a pattern inherited by early therian ancestors. Only with the development of a complex allantoic placenta with secretory properties permitting the suspension of oestrous allowed a prolongation of the gestation period and leading to more developed neonates.

Consequently, the production of advanced young is considered to be a secondary evolutionary development in mammals (Case 1978a) which evolved relatively late in mammalian history. The reproductive features of *Rhynchocyon* would fit into this hypothesis as it is the most ancient genus (e.g. Tabuce et al. 2008) within the order Macroscelidea. It has the shortest gestation time and has the least precocial offspring among the other sengis.

On the other hand, long pregnancy is considered an ancestral trait in Bathyergidae Sumbera et al. 2003). The "primitive" taxa Ctenodactylidae and Pedetidae have relatively longer developmental times than other sciurognaths and even produce precocial young (Burda 1989a). Conseuqently, longer developmental times with precocial young are considered a derived pattern in hystricognath rodents (Sumbera et al. 2003) and only in small hystricognaths altricial young are born after a long pregnancy.

The majority of extant mammal species produce altricial offspring (Case 1978a). But the so-called altricial/precocial dichotomy exists within several taxa suggesting that the ancestral altricial developmental mode (Hopson 1972, Case 1978a) has been modified towards precociality more than once during mammalian evolution. That means that mammalian development patterns do not map onto continuous changes in life history patterns and a shift from altricity to precociality and vice versa during the evolution is very probable.

The cost of precociality is a low intrinsic rate of natural increase and appears to depend on body size, such that differences in potential population growth rates are greatest in small-bodied mammals. This might explain why only a few species of small mammals produce precocial young (Hennemann 1984). *Rhynchocyon* has the shortest gestation length but produces mostly one semi-precocial offspring in contrast to *Elephantulus* with up to 4 precocial offspring but much longer pregnancy. In order to increase the reproductive rate the Caviidae were able to shorten their developmental time (postnatal phase) (Burda 1989a) and evolved extreme precociality (Kraus et al. 2005). These authors concluded for caviomorph rodents that phylogenetic inertia is considered to constrain a shift to altricity which could also be true for sengis.

1.5.2.3 Precocial and altricial strategies
Neal (1986) found in his analysis of reproductive characters of African small mammals including *Elephantulus,* that the altricial/precocial dichotomy does not fit the prediction of r- and K-selection theory but that these 2 main strategies have a mixture of r- and K-selected characteristics although species with altricial young tend to be more r-selected than species with precocial young. Consequently, reproductive effort (the proportion of total energy budget allocated to reproduction) should be a measure of an animal's position along a continuum of "r- or K-selected" strategies (Millar 1977).

1.5.3 Gestation length

Generally, it is accepted that gestation length is correlated with the degree of development of the young at birth and resulting in increased development of the central nervous system (Sacher and Staffeldt 1974). Martin and Mac Larnon (1985) clearly distinguish altricial and precocial mammals by gestation period relative to maternal weight despite of some overlap between large-bodied altricial mammals and certain small-bodied precocial mammals. In contrast, Case (1978a) found no consistent correlation between adult body size and the degree of maturity at birth. *Rhynchocyon sp.* is the largest sengi but the least precocial. The question remains if neonatal developmental status in relation to adult body size and gestation periods are appropriate allometric patterns for phylogenetic research.

However, costs and benefits of prenatal and postnatal development must be balanced no matter what the developmental status of the neonate is. On the other hand, pregnancy is energetically much less demanding than lactation and provides the offspring with a high degree of protection and environmental stability (Millar 1977). Weir and Rowlands (1973) expressed the difficulty determining causal directions with the particular example of gestation length as follows: "it is obviously impossible to tell whether the gestation length was the fixed parameter and the environmental cues existing at the time were incorporated into the mating season stimuli, or whether the gestation length was the variable upon which selection acted."

1.5.4 Parental care

Fathers do not seem to be involved in infant care in captive *Rhynchocyon petersi* and no interaction was observed during the in-nest period. Only casual interaction occured after the infants emerged from the nest. It was not clear whether there was a male's participation in the nest building (Baker et al. 2005). Apparently, in the course of the evolution of the mammary glands the role of the father in many mammals was reduced as is true also for sengis.

With regard to the high neonate birth mass in macroscelids in comparison to the mother's weight (ca. 50% in a twin litter) it is evident that the investment during pregnancy must have reached the limit of a female's physical capacities. This investment favours the absentee maternal care system (Rathbun 1979a, Rathbun and Rathbun 2006a) and neonates are left alone for a long period of time shortly after birth already (Sauer 1971 and 1973, Baker et al. 2005). This system could have evolved to ease the female's burden of rearing young after this long pregnancy.

Male *E. myurus* (Ribble and Perrin 2005) and *Petrodromus* (Rathbun 1979a, FitzGibbon 1995) have never been observed in the same vicinity of females with offspring. Sengis will not directly defend their young but predation is avoided in *E. rufescens* and *R. chrysopygus* partly by running away (Rathbun 1979a) which reduces the risk of predation also in mobile precocial young.

Sauer (1972) emphazised the importance of a secure birth site on which the *Macroscelides* mother can rely on rather than helping her offspring in the case of danger. He also observed a large distance between the usual home range of the mother where she searches for food and the site the offspring are hiding.

1.5.4.1 Lactation

In the course of evolution reproductive hormones allowed the formation of incubation areas for further development of foetuses and further, in some mammals also glandular secretions. Lactation is considered a crucial factor in the evolution of mammalian parental investment and reproductive strategies (Eisenberg 1981). The lactation period, while maternal feeding is crucial for the surviving of the young, is shorter than the weaning time.

1.5.4.2 Feeding intervals

Female *Macroscelides* (Sauer and Sauer 1972) and *R. petersi* (Baker et al. 2005) return only once a night to their newborn young to nurse them. Rathbun (pers. comm.) speculated that this absentee maternal care system (Sauer 1973) may relate to the relatively large number of teats in relation to small litter size. Within 80 minutes *Macroscelides* young were nursed 6 times which implies a large amount of milk. Further, relatively large offspring could survive longer between feedings so that the female could feed her young infrequently and spend more time feeding (Millar 1977).

The guinea pig belongs to the most precocial species among small mammals but compared with sengis, the female invests much time in litters of few young (Kraus et al. 2005). The suckling interval of caviomorph rodents varies quite widely from about 2 to 35 minutes (Alderton 1996). Species which produce their offspring in underground burrows such as the degus (*Octodonta spec.*) spend longer suckling them than e.g. the caviids, which remain vulnerable in the open.

1.5.5 Monogamy

Monogamy in sengis is considered a unique pattern which is shared by the whole clade (Ribble and Perrin 2005). There is no evidence that monogamy evolved in response to the need for paternal care and evolved more often in the absence of paternal care than in its presence (Komers and Brotherton 1997) and the authors conclude that the costs and benefits of monopolizing solitary females appear to be the primary issues in the evolution of monogamy in mammals.

Monogamy in sengis may also be related to testes size as observed in other mammals (Woodall and Skinner 1988) and lack of sexual dimorphism e.g. in *R. chrysopygus* (FitzGibbon 1995).

1.5.5.1 Monogamy and parental care

Zeveloff and Boyce (1980) generally found that species with low maternal investment in the young at birth tend to be monogamous. Primates (*Lemur, Tarsius, Cercopithecus*), lagomorphs (*Ochotona, Caprolagus*), rodents (*Heterocephalus, Dolichotis*), carnivores (*Fossa*) and some ungulates are monogamous but lack paternal care.

The question arises how paternal care can be defined. Male and female sengis do not have much direct contact which is defined as uniparental monogamy (Rathbun and Rathbun 2006b) which is also known for the dik-dik (Brotherton and Rhodes 1996). Difficulty is experienced in deciding what the benefits of monogamous parents for offspring survival are, if there is neither direct male care nor protection from infanticide or predators. Trail maintenance activities may be one of the benefits providing easy access to the territory for foraging and predator escape (Rathbun 1979a) but this trait is not generally present in all sengis (Ribble and Perrin 2005).

1.5.5 2 Mate guarding

The dik-dik model (*Madoqua kirkii*, Brotherton and Rhodes 1996) was used to discuss mate guarding (Ribble and Perrin 2005) as the best explanation for social monogamy in sengis. They found that dik-diks and sengis share many life history traits including small litter size and highly precocious young. Male mate-guarding seems to be responsible for monogamy in *E. myurus* (Ribble and Perrin 2005) and *R. chrysopygus* (FitzGibbon 1995). Lumpkin and Koontz (1986) suggested that mate guarding in view of the short duration of oestrous in sengis might be particularly adaptive which has led to monogamy (Brotherton and Manseri 1997). Male mate-guarding has recently been proposed as the principal adaptive factor in the evolution of all mammalian social monogamy with parental care of young likely being secondarily derived (Ribble 2003).

But unlike antelopes sengis do not exhibit strong pair bonds (Rathbun and Rathbun 2006a). In *M. proboscideus* even so few intra-pair interactions occur that Sauer (1973) characterized the species as solitary, but strongly promiscuous. He still supports a latent pair-bond as neighbouring partners will breed again in the next oestrous cycle. This strategy shows that male sengis are not forced into obligate monogamy such as dik-diks which have to spend much of their time with their mate to prevent them from leaving the pair's territory (Komers 1996). But even in strictly monogamous species with no evidence of extra pair paternity such as the dik-dik, males did attempt to obtain extra-pair copulations whilst females did not attempt to do so (Brotherton et al. 1997).

1.5.5.3 Monogamy and precociality

The tendency for monogamy related to short gestation periods as postulated by Zeveloff and Boyce (1980) however, is controversial to the relatively long gestation period in sengis. Zeveloff and Boyce (1980) also found that species with altricial young tend to be monogamous and those with precocial young tend to be polygynous. In contrast to this theory almost all sengis are precocial and monogamous whilst the aardvark produces less developed young and lives in a polygamous mating system (Shoshani et al. 1988). However, analysing data from insects, birds and mammals regarding the breeding success of males showed that the differences between monogamous and polygynous species are smaller than expected (Clutton-Brock 1988), especially when offspring survival is included in measures of male success.

1.5.6 Juvenile mortality

Risk-sensitive foraging behaviour is rather one of the factors to the benefit of both offspring and mother as this decreases the mortality rate as observed in *Macroscelides* (Lawes and Perrin 1995). The authors noted a substantial behavioural flexibility where "natural selection has acted strongly in shaping the foraging behaviour". This means that the probability of energy shortfall or starvation is minimized and risk-prone foraging is highly unlikely which would attain a positive effect on the successful rearing of young. Van der Horst (1946) assumed an infant mortality of about 55% for wild *E. myurus jamesoni* which in his view is low. In captive *Macroscelides* two third survived longer than a month (Olbricht et al. 2006) and about 80% of captive *E. rufescens* (Rathbun et al. 1981).

Hennemann (1984) states that the evolutionary "decision" to produce precocial young may allow increased survivorship of the young in response e.g. to predation and this agrees with Case´s notion (1978a) that the relative intensity of predation on the young compared with that on the adult appears to be most important in determining the optimum size of the neonatus and its maturity at birth.

1.5.7 Litter size

The question arises whether adult females are breeding at or close to the theoretical limit of their capacity to produce and rear young. Partridge (1989) postulates "that number of young produced per litter is often not the maximum number that a female can produce in a breeding event but a set of trade-offs between the mothers costs of reproduction and her future reproductive potential and between number and size of young". Because precocial young would be at the maximum mass that the adult female can carry selection for short nestling periods and concomitant longer gestation periods would lead to small litter size (Viljoen and du Toit 1985). It could also be a limiting factor, that speed is required to avoid predators in the sense that any selection against to heavy females during pregnancy would favour a smaller litter size (Millar 1977) and he concludes that litter size appears to be the adaptive variable of mammalian reproduction. Millar also found that another nonenergy factor that could favour small litters is the ability of a female with precocial young to keep track of her offspring.

1.5.7.1 Correlation between teat number and litter size

Phylogeny only influences litter size through homology in that a number of related taxa may have a similar litter size because a common ancestor became adapted, e.g. to a stable climax environment (Perrin 1986). Mammary number and litter size are not similar among sengis but in all genera litter size is much larger than teat number. However, there might be homology to other afrotheres. Similar to *Rhynchocyon* and *Petrodromus* mostly one young is born in the aardvark (*Orycteropus afer*, Shoshani et al. 1988), the Florida manatee (*Trichechus manatus*, Reynolds and Odell 1991) and Grant's Golden mole (*Eremitalpa granti*, Perrin and Fielden 1999), all females have 4 teats. Teat number and litter size within the Tenrecidae varies remarkably (Puschmann 2004).

A large number of teats may be useful in connection with big litter sizes and the intensive care of dependent young. Among rodents not all species fit into the scheme of the "one-half-rule". Comparable to sengis some similar characteristics are found in 2 other precocial species: the rock hyrax (*Procavia capensis*, Olds and Shoshani 1982) and a rodent, (*Chinchilla lanigera*, Anderson and Sinha 1972) which have 6 teats but only 2 - 3 offspring. The latter authors argue about a correlation between mammary number and litter size. The naked mole rat (*Heterocephalus glaber*) is an altricial rodent but it is an exception to the "one-half rule" (Sherman et al. 1990). Its litter size doubles teat number.

Pearl (1913 a, b) uses the term "teleological necessity" for a compensating increase in the number of mammae when the number of young born in a litter increases. But this is a rather philosophical attempt and nobody can´t imply that evolution is a directed process. In contrats, the additive genetic variance appears to be less for mammary number than for litter size (Gilbert 1986). Polar bears (*Ursus maritimus*) normally have 4 mammae but up to 6 functional mammae are reported with less than 2 cubs as average litter size (Derocher 1990). Some of the supernumerary teats were suckled. In comparison to the Brown bear (*Ursus arctos*) with 6 teats and 2 cubs per litter Derocher suggested that reduction in litter size and the normal number of mammae may be related to the difficulty of raising cubs in the harsh environment of polar bears which agrees with Erickson (1960) that supernumerary teats are a throwback to an ancestral condition. The Zambian common mole rat (*Cryptomys sp.*) is an altricial rodent which has 3 teat pairs, the mean litter size is 2 but can be up to 6 young. In a litter of 1 or 2 offspring only the pectoral teats are suckled. If more young are born, the axial and inguinal teats are used too (Burda pers. comm.) which is in contrast to hyraxes where the inguinal teats are mostly preferred (Hoeck 1977, Fischer 1986). Nevertheless, given that the conception of phylogenetic inertia (many teats as ancestral character) applies it would be interesting to investigate if unsuckled mammae would disappear over evolutionary time as observed in domestic sows (Kim et al. 2001).

However, the one-half rule does not apply for sengis. A phylogenetic interpretation of the large number of teats in sengis is somewhat hampered by uncertainty concerning their relationships to other mammalian orders. I infer that the large number of teats might be a relict from ancestral times and is explained by phylogenetic inertia. Secondly, a larger number of teats provide a larger amount of milk which may benefit precocity. Especially the Macroscelidinae are small mammals with small sized teats, barely visible even while lactating. It may be speculated that in these highly cursorial mammal the amount of milk needed is distributed among several teats rather than in a few udder-like bigger teats.

1.5.7.2 Limits of litter size

Litter size may be limited by one or more of several factors, depending on the habits of the animal. Litter size in mammals may be limited by the ability of females to carry large foetal biomass and consequently, any selection against heavy females during pregnancy would favor a smaller litter size (Millar 1977).

Mammary number in the Macroscelidea apparently does not limit litter size and subsequently reproductive success as e.g. in marsupials (Ward 1998) where teat clinging is essential of the survival of neonates. In most mammals the reduction of litter size at birth is the result of post-implantation loss such as in some diprotodont marsupials which typically ovulate more eggs and, in contrast to sengis, supernumerary neonates are produced (Ward 1998). In comparison to marsupials where litter size is restricted by the number of functional teats he suggested that restriction in teat number may be analogous to the restricted implantation sites of the sengis. Here the question remains unanswered why in sengis with mostly twin litters we find 6 functional mammae and despite of the restricted area of the implantation site supernumerary litters can be produced as reported for *M. proboscideus* and *E. rozeti*.

Given this conservatism of a species-typical invariant mammary number, excluding exceptions, there must be other mechanisms compensating for constantly increasing litter numbers. Some altricial rodents are known to adjust litter size by infanticide during early lactation (Perrigo 1987) whilst this has never been observed in the precocious guinea pigs (Künkele 2000). As a result, a more adequate mechanism to reduce litter size is probably fetal resorption during gestation (Peaker and Taylor 1997). Precociality as mentioned earlier also leads to smaller litters.

1.5.8 Seasonality of breeding periods

Although macroscelids have populated various ecological niches distinct reproductive seasonality is not found. The number of litters produced per year and the survival of young may be strongly limited by the duration of the breeding season (Case 1978b). Case (1978a) also suggested that the attainment of endothermy in the ancestors of sengis must eventually have extended the reproductive season of early macroscelids because it enabled these organisms to physically buffer seasonal changes to a greater degree.

The combination of a long breeding season with a smaller litter size enables some tropical rodents to spread nutritional demands of breeding over a larger proportion of the year than is possible in temperate climates (Viljoen and du Toit 1985). *E. rupestris* is sexually active throughout the year assuming that they are less dependent upon rain and fog for reproduction than are e.g. the rodent species of the same area (Withers 1983). Neal (1986) noted that species bearing precocial young have relatively aseasonal breeding patterns even in very seasonal environments compared to species with altricial young because precocial species may have lower energy requirements. This would give precocial species a better option to reproduce aseasonally because maximum energy expenditure during lactation is not as extreme as in rodents with altricial young (Trillmich 2000). *Macroscelides* adopted an omnivorous diet (Kerley 1995, Bernard et al. 1996) to deal with energy constraints which consequently allows reproduction throughout the year.

Natural selection may favour rapid growth rates in highly seasonal environments (Zeveloff and Boyce 1986) which benefits the evolvement of precociality. The genus *Elephantulus* is physiologically and behaviourally adapted to arid and semi-arid regions of low productivity. It cannot compete against species bearing more altricial young in more productive habitats and Neal (1995) concluded that the precocial habit may be one reason why this genus is restricted to such environments. But reproductive activity is not only influenced by the environment but also by the organism's own characteristics (Neal 1986). On the other hand, sengis exhibit a broad behavioural flexibility along with morphological and physiological features shown in observations about thermoregulation (Roxburgh and Perrin 1994, Perrin 1995) including torpor (Lovegrove et al. 2001a, b), feeding behaviour (Kerley 1995, Lawes and Perrin 1995) and general natural history (Sauer and Sauer 1971, 1972) to cope with environmental stochasticity (Lawes and Perrin 1995).

1.5.8.1 Male capacities

Woodall and Skinner (1989) described relatively small testes but extra-gonadal sperm reserves in *E. myurus jamesoni* compared to many similar-sized mammals. They found that this could be compensated by the active spermatogenesis throughout the year which is subsequently leading to continuous breeding during an extended breeding season. In disagreement to Stoch (1954) and van der Horst (1946) there is no cessation of breeding. Woodall and Skinner (1989) found active spermatogenesis in male *E. myurus* throughout the year despite of significant reductions in testis and prostate size outside the main breeding season. In contrast to other precocial mammals such as the chinchilla and the plains viscacha (*Lagostomus maximus*) where the males are always fertile and only the females are seasonal (Weir and Rowlands 1973), in sengis there is no complete cessation of breeding.

1.5.9 Post-partum oestrous

Inhibiting oestrous during lactation would not benefit the improvement of the reproductive rate for species with well developed young and short lactation (Weir and Rowlands 1973). There is no lactation anoestrous in *E. myurus jamesoni* (van der Horst 1946) but lactation following one pregnancy overlaps with the subsequent pregnancy (McKerrow 1954) which increases the reproductive effort. Similar observations on rodents revealed that post-partum oestrous might be an indicator of high reproductive effort (Gilbert 1986).

1.5.10 Poly-ovulation

Natural poly-ovulation is a rare process in mammals (Weir and Rowlands 1973). In *E. myurus* 120 eggs may be liberated from the 2 ovaries at each ovulation (van der Horst and Gillman 1941a), a number which is exceeded only by a hystricomorph rodent, the plains viscacha (*Lagostomus maximus*) which produces 200-800 eggs (Weir 1971). This species shows some similarities to sengis, as it has precocial young, breeds throughout the year, lives in pairs and is poly-oestrous. Also, a female aardvark with one foetus and 5 more corpora lutea was found (Shoshani et al. 1988). For *Tenrec ecaudatus* the number of liberated eggs is not known but the litter of 31 young (Louwman 1973) is supposed to be the biggest among all the other Afrotheria. Poly-ovulation in tenrecs and sengis may represent some relict mechanism in ovulation procedures (Weir and Rowlands 1973).

At first glance, it appears that some successful reproductive traits evolved in very different taxa due to environmental constraints. On the other hand, variations in the ovulation rate among sengis appear to be unrelated to environmental factors as there is an overlap of the distribution of many oligo-ovulating and poly-ovulating sengi species (Corbet and Hanks 1968).

1.5.10.1 Female capacities

The reduction in the number of eggs from 60 to 1 (*E. myurus*) in each uterine horn is achieved at the latest four-cell stage and before implantation (van der Horst and Gillman 1941a). There is only one very small and well circumscribed area in each uterine horn, which enables the successful implantation of a single embryo at a time (van der Horst 1942; van der Horst and Gillman 1942c). Consequently, the final litter size is much likely to be determined entirely by the restricted implantation sites (Tripp 1971). This embryo chamber is a consistent feature of placentation in sengis that may partially account for the small litter size (Oduor-Okelo et al. 2004). *E. rozeti* apparently is exceptional where 2 embryos are imbedded in the caudal end of each horn. Triplets were also found in *M. proboscideus* (Olbricht et al. 2006).

This reproductive potential including potentially supernumerary young and poly-ovulation could help to shed some light on this question. Typical prey species stand under a strong pressure of natural selection and must be prepared for the occasional extreme rather than for the more frequent mean litter size. Sherman et al. (1999) suggested that extreme and variable litter sizes may be related to unpredictable food availability which agrees with Pearl (1913a) who proposes an evolutionary "factor of safety" to achieve flexibility with environmental cues. Subsequently, a larger number of teats would support this flexibility and leaves space for further development within the evolutionary framework.

1.5.11 Longevity and fecundity

Two determining factors for the fecundity of individuals are lifespan and inter-litter intervals (Promislow and Harvey 1990). One strategy to produce enough offspring to balance the death rate is to extend the length of the reproduction period (Neal 1986). In fact, continuous breeding with short-inter-birth intervals and a long lifespan (Olbricht 2007) reveal the high reproductive potential of sengis despite of their low reproductive rate (small litter size, long gestation periods).

Sengis have a longer lifespan as one may expect from species of that size (Jones 1985, Weigl 2005) which is linked with an extremely high fecundity rate until the end of their life (Olbricht 2007). This is also known for Zambian mole rats (*Fukomys anselli*), small, altricial subterranian rodents. Data from captive sengis reveal that lifespan of specimens observed in the wild or earlier studies in captivity are exceeded by far as a result of improved husbandry (Jones 1982, Weigl 2005). Van der Horst (1946) had suggested that female *E. myurus* and consequently all other small sengis may have only 3 pregnancies during their life time but improvement in breeding also led to high fecundity rates which was found in captive *Macroscelides* (Olbricht 2007) where a female had 23 litters with 46 offspring during her lifespan of 5 years and 2 months, among those the unusual birth of triplets. She had the last litter 3 months before death.

Additionally, sengis have a number of behavioural and ecological traits that may represent adaptations to improved survivorship such as predator avoidance by young (Sauer 1972) and by adults (Rathbun 1979b), risk-.sensitive foraging (Lawes and Perrin 1995), defending a territory (Ribble and Perrin 2005), thermoregulation (Perrin 1995) and a varied diet (Bernard et al. 1996). The "K-selected" traits of sengis require an effective quality breeding performance over a long life time. This is in line with Harvey and Zammuto (1985) and Promislow and Harvey (1990) who support an evolutionary link between low mortality rates, delayed reproduction and long lifespans.

1.6 Conclusions

The reproductive characteristics of the Macroscelidea are unusual as suggested by many authors (e.g. Neal 1995). Ancestral sengis were presumably not smaller in size than extant species (Evans 1942, Patterson 1965, Simons et al. 1991) but probably had altricial offspring, as is suggested for other ancestral small mammals too (Derrickson 1992). Whereas the Rhynchocyoninae maintained ancestral features such as shorter gestation time, leading to less mature offspring and longer post-natal development, the Macroscelidinae extended pregnancy which is linked to precociality and shortening of the post-natal phase. The post-natal development of body metrics is presented in chapter 5 and 6.

Life history traits of sengis such as fecundity and and lifespan as well as reproductive patterns such as litter size, post-partum oestrus, neonatal development and mammary number reveal a high reproductive potential and could balance the cost for the evolvement of precociality. These patterns may also have evolved in general to achieve flexibility with environmental cues which makes it difficult to determine causative factors during their phylogeny. As a result, it cannot be fully clarified whether precociality in sengis is a plesiomorphic trait.

Rhynchocyon and *Petrodromus* have fewer teats than *Elephantulus* and *Macroscelides*. If I assume in a first hypothesis that phylogenetic inertia is the reason for this generally large mammary number relative to litter size and this is an ancient trait, so all the other physiological and behavioural traits analyzed here are adaptations to the sometimes extreme living conditions of the macroscelids. There has possibly never been a necessity to reduce the number of teats because behavioural adaptations could even derive from it as an advantage. If, for example, the maternal absentee care system can be considered as the main reason for upholding the large mammary number there could be a direct advantage of this strategy of postnatal care as it eased the dam´s work load of rearing young. In most macroscelids the highly precocial young require for a few days only an obligatory milk supply which consequently increases the fitness and the energy supply for the subsequent litter. A short but efficient milk supply is required which is secured by the large teat number but on the other hand, 6 functioning mammae are quite energy requiring. A lot of energy and effort is required but for a short period of time only and precociality is favourable to reduce the time of dependence of the offspring. Consequently, breeding success is not negatively influenced by parental care of young of different litters but is increased by short interbirth-intervals and aseasonal breeding. In domestic guinea pigs (*Cavia porcellus*) the peak of energy demand is low relative to that of altricial rodents and extended pregnancy is linked to a high efficiency of energy conversion into offspring (Künkele 2000). These patterns in reproductive energetics led to the speculaton that the precocial strategy might be ecologically robust and enables wild guinea pigs to reproduce year-round, even under harsh conditions. With regard to the distribution of sengis in dry habitats this aspect could also apply on macroscelids. Additionally, traits such as precociality, absentee maternal care system and many teats together with the sengis´ life style help to compensate that sengis as typical small prey species are vulnerable to predation.

Another hypothesis becomes important with respect to the most ancient genus *Rhynchocyon* which is living in dense forest. On the base of its life history I could assume that the ancient feature was 4 teats and not 6. The following conclusions would support this hypothesis. The more specialized *Elephantulus*-species and *Macroscelides* are inhabitants of open habitats. The potential to produce more offspring than the usual number of 1 or 2 is clearly evident in *Macroscelides* as well as in *Elephantulus rozeti*. As a result, an increase of mammary number as an adaptive evolutionary trait could be possible. Six teats might then be considered as a derived trait from the original 4 teats. However, it will not explain why *Petrodromus* also has only 4 teats.

Poly-ovulation occurring in several species is another indicator for a future potential increase of the reproductive potential. Although oligo-ovulation occurs also in *Elephantulus*-species, *Petrodromus* and *Rhynchocyon* do not seem to be provided with the same reproductive potential as the genera *Elephantulus* and *Macroscelides*. The gestation period is shortest in *R. chrysopygus* (42d) and longest in *E. rozeti* (75d). The highly precocial young of the latter are born in a harsh and dry environment and do not build nests. In contrast to *Rhynchocyon* neonates, who find the safety of a nest in dense cover, they need to develop more rapidly as they are more exposed to life threatening risks.

The social monogamy found in the Macroscelidea across species and across different habitats may be the consequence of the conservation and interaction of a suite of uniquely derived traits (Rathbun and Rathbun 2006a).

Continuous breeding occurs in all sengis and implies post-partum oestrous. This could be one of the reasons why the archaic *Rhynchocyon* could survive for 20 millions years in which time their basic anatomy and presumably their habits have changed very little (Kingdon 1990).

It was to examine if these characters could be a source of phylogenetic information within the Macroscelidea and if so, that they might even give support to the Afrotheria clade.

With respect to the macroseclids relationship to the Afrotheria it seems that at least neonatal development patterns are not a good tool to detect phylogenetic interrelations as "they do not appear to result from one continous axis of selection" (Derrickson 1992). In their review on afrotherian origins Robinson and Seiffert (2004) came to the inference that afrotherians, except aardvarks and paengulates are unanimous and that their phylogenetic affinities lay with a variety of other non-paengulate placental orders. Subsequently, for different afrotherian species independently from phylogenetic relationships interspecific correlations may have evolved.

Given their ancient association with the Afrotheria and their clinging to their life history traits phylogenetic inertia seems to play an important role why sengis did not evolve more specializations e.g. in their social organization (Rathbun and Rathbun 2006a). Nevertheless, sengis and probably other afrotherians too, developed a large number of adaptations to compensate for their legacy. This agrees with Robinson and Seiffert (2004) who found it very difficult to determine which taxa might preserve the morphological apomorphies that were present in the afrotherian ancestor and which taxa might have had their afrotherian synapomorphies erased. The value of reproductive parameters alone for phylogenetic research seems to be limited but together with other morphological and molecular patterns it is a valuable source of additional information.

CHAPTER 2

2. THE TOPOGRAPHIC POSITION OF THE PENIS IN SENGIS, AND COMMENTS ON PENIS TOPOLOGY IN TESTICOND MAMMALS

(data were partially obtained and computed in collaboration with Dr. T. William Stanley, Field Museum of Natural History, 1400 South Lake Shore Drive, Chicago, IL 60605, USA)

2.1 Abstract

In order to determine if the position of the penis along the longitudinal axis of the body, relative to other morphological landmarks is phylogenetically informative 44 wild born and captive specimens of all four sengi genera were measured and analyzed. The position of the penis is of utility in distinguishing between the genera *Petrodromus* and *Macroscelides*, but not between any other genera of sengi, supporting recent taxonomic conclusions regarding the relationship of these two taxa. In juvenile short-eared sengis (*Macroscelides proboscideus*) the position of the penis was significantly closer to the anus than in the adults of the same species.

2.2 Introduction

Testicular descent (testicondy) with respect to a certain variability identifies sengis as testicond afrotheres (Werdelin and Nilsonne 1999). While both historical and recent studies have focused on the natural history of the Afrotheria, limited or conflicting information is available on male sexual organs of the Macroscelidea. Woodall (1995a) reviewed the sengi male reproductive system and discussed the phylogeny of the order. Stoch (1954) described the male reproductive tract without discussing the penis. Earlier studies (Peters 1852, Carlsson 1909, Kaudern 1910) of the anatomy and histology of the penis in sengis gave accurate descriptions; however Woodall (1995b) later maintained that the results for the subject species of these studies are not representative of all genera of the order. Extending earlier studies, Woodall (1995b) and Woodall and Fitzgibbon (1995) conducted detailed examinations of the variation in morphology and histology of the penis for the various species of sengi, and found that the structure of the penis is one of the distinctive features within the group. Douady et al. (2003) investigated the amount of congruence between phylogenies based on morphology of sexual organs and those generated from analysis of molecular evidence, in a study that found that the genus *Elephantulus* to be diphyletic. The data also revealed close taxonomic relationship between *Petrodromus tetradactylus* and *Elephantulus rozeti*.

2.2.1 Sengis as testicond afrotheres

In sengis, the testes are just caudal to the kidneys, similar to other testicond mammals (Woodall and Skinner 1989, Woodall 1995a). The penis is very long and runs anteriorally under the abdominal skin – a feature not found in other mammals (Woodall 1995a, Frey 1994a). It emerges just caudal to the sternum and takes up to 75% of the body length in *E. rozeti* and up to 50% in *P. tetradactylus* (Frey 1994a). This long body of the penis presumably is the reason why the genital opening is situated far anteriorally. Thus a relation between penis length and anus-penis distance can be expected. However, another afrothere has evolved a different strategy: While sengis do not have a baculum, the Tailless tenrec (*Tenrec ecaudatus*), with a penis that takes up to 75% of the body length, is the only afrothere which has one (Frey 1994a). That two groups of mammals with similar structure in male phalli differ in possession of an os penis begs further investigation.

2.1.2 Taxonomic tools

Penial morphology provided valuable information on the systematics of rodents (Altuna and Lessa 1985, Lidicker and Brylski 1987) and viverrids (Wemmer et al. 1983). Morphological features of the penis, as well as spermatozoa, are of use in taxonomic distinction among genera of the order Macroscelidea. But information on penis-anus distance in mammals in general is scarce. Difference in this has been considered as a feature of some taxonomic significance in the order Hyracoidea, (Coetzee 1966, Glover and Sale 1968) and special attention was paid in this study to this order which is also a member of the Afrotheria.

While various publications regarding sengis mentioned the unusual position of the penis (Sauer 1973; Woodall 1995a), quantitative data with detailed measurements of its location relative to other landmarks such as the anus, are lacking, and the relevance of the anus-penis distance in taxonomy is unclear.

2.1.3 Aim of the study

Here we examine whether the topographic position of the penis using specimens of all four genera
of the order, to determine whether the penis-anus-distance is phylogenetically informative within the order Macroscelidea. Another aspect of the paper elucidates the question why particular morphological structures may have evolved. With respect to differences in the penial morphology and topology, Frey (1994a, b) found relationship between the length of the penis, mode of locomotion and copulatory posture in mammals. The results are discussed within the context of male reproductive features in other testicond mammals, especially among the Afrotheria.

2.2 Material and Methods

2.2.1 Material

44 specimens of eight different sengi species representing all four extant genera within the order, and from seven different locations were examined and measured (Table 2.1). Some of the specimens were wild caught, and others were born and raised in captivity. The age of captive animals was recorded when known. All specimens were preserved in fluid (either formalin and/or ethanol).

2.2.2 Methods

Measurements were taken using calliper and ruler, as follows: the specimen was placed on its back, and the penis (not expressed from the prepuce) was positioned against the abdomen along the longitudinal axis of the specimen with the distal tip of the prepuce oriented anteriorally. Straight-line distances were measured from the anterior tip of the lower lip of the mouth to the distal end of the prepuce (PL) and from the distal end of the prepuce to the anus (PA) (Fig 2.1).

In order to assess whether the penis position differed between adult and juvenile animals, a sample of 15 male *M. proboscideus* of known age was separated into adult (> 3 months old, n = 8) and juvenile (< 3 months old; n = 7). Because the preparation of the specimen may present the potential for variation among measurements, the ratio of the distance of anus to penis/ anus to lower lip was used to compensate for the measurement error.

Steiner and Raczynski (1976) stress out measurement errors, and urge repetition of the measurement where a single or only a few specimens are available. Therefore, some of the extreme accounts were double checked. The averages of the two measurements were subjected to further analyses.

The total length (AL) is the sum of these two measurements (PL + PA). Penis location was determined by computing the percentage of the distance between anus and prepuce relative to the total length (AL/PA x 100). While the quality of the material differed due to preservation methods as well as subsequent treatment (such as removal of the skull or ventral openings), the use of ratios involving penis placement relative to anus and total length allowed suitable comparisons among different specimens.

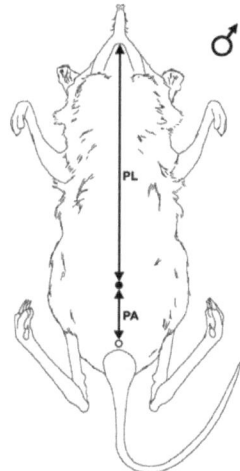

Fig. 2.1 Measurement methodology to determine penis location (schematic drawing by R. Banasiak)

2.2.2.1 Hyracoidea

Data from the literature for the Hyracoidea were computed with the same methods as used previously for sengis enabling thus comparison with the Macroscelidea. From all these accounts for body lengths (BL) of *Procavia capensis* (Olds and Shoshani 1982) and *Heterohyrax brucei* (Barry and Shoshani 2000), mean values were computed for each of the two species. These means were set in relation to the distance anus-penis of each species (Coetzee 1966, Glover and Sale 1968). The resulting percentage is comparable to the percentage which was computed for sengis.

2.2.2.2 Statistical methods

Standard descriptive statistics (mean, range, and standard deviation) were calculated (Table 2.2). The data for each genus were subjected to the Shapiro-Wilk W test to gauge whether the distribution of each set was normal. One-way analyses of variances (ANOVA) were used to determine whether the age had a significant effect on relative position of the penis in *Macroscelides*. To test for and assess significant taxonomic variation a one way ANOVA (effect = genus) and the post-hoc Sheffé-test to identify characters that differed significantly among groups. All analyses were conducted using Systat version 11 or SPSS 15.0.

2.3 Results

2.3.1 Sengi measurements

Among all the sengi species measured, including both juvenile and adult individuals the position of the penis ranged from 24 to 41% along the longitudinal axis relative to the distance from anus to lower lip. The measurements for each individual is presented in Table 2.1, the mean values of penis position relative to body length for the entire genus are given in Table 2.2.

The relative position of the penis was 30% (24 to 35%) of the total length in juvenile *M. proboscideus* and 37% (34 to 41%) in adults with an average of 34% for all specimens of this species. A one-way analysis of variance demonstrated for *M. proboscideus* that the distal tip of the penis was significantly closer to the anus in juveniles than it was in adult specimens ($F = 8.4$; $p < 0.05$). Figure 2.2 illustrates the penis position of a juvenile and adult *M. proboscideus*.

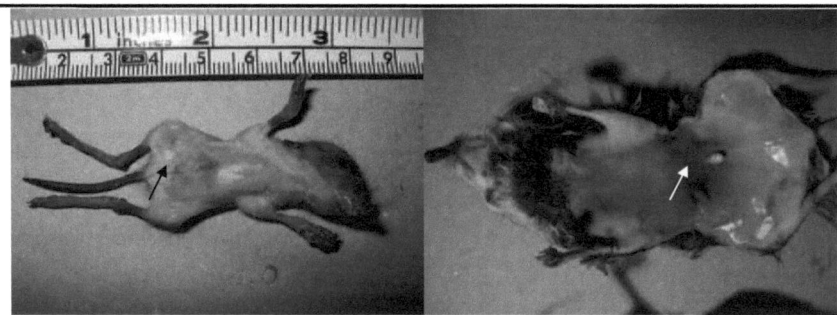

Fig. 2.2 Penis location in *M. proboscideus*, left: stillborn (M205007A), right: 6 months old (M240048)

Table 2.1. Measurements for individual male sengi specimens. The distance between the anus and penis (PA), between penis and the lower lip (PL), and total (AL; based on the sum of the first two measurements) are in mm. The penis location is represented as AL/PA X 100. Known juvenile specimens (< 3 months old) are marked with grey. ZFMK: Museum Alexander Koenig, Bonn, Germany; FMNH: Field Museum of Natural History, Chicago, USA; RMCA: Royal Museum of Central Africa, Tervuren, Belgium; German zoos include Zoo Bernburg, Zoo Wuppertal, Zoo Halle, and Zoo Erfurt.

Species	locality	PA	PL	AL	Penis location	Museum	Catalogue #
Petrodromus:							
P. tetradactylus	Djabir, Zaire	44	135	162	24%	RMCA	AI097M403
P. tetradactylus	Tanzania	48	137	185	26%	FMNH	163413
P. tetradactylus	Tanzania	37	104	141	26%	FMNH	193245
P. tetradactylus	Tanzania	48	123	171	28%	FMNH	151462
P. tetradactylus	Tanzania	36	83	119	30%	FMNH	169665
P. tetradactylus	Tanzania	49	86	135	36%	FMNH	193247
P. tetradactylus	Tanzania	55	99	154	36%	FMNH	151460
Elephantulus:							
E. rufescens	Sudan	31	75	106	29%	FMNH	144103
E. rufescens	Tanzania	28	67	95	29%	FMNH	193297
E. rufescens	Tanzania	31	74	105	30%	FMNH	193391
E. rufescens	Sudan	33	54	87	38%	FMNH	144070
E. rufescens	Sudan	37	59	96	39%	FMNH	144099
E. rufescens	Sudan	32	55	87	37%	FMNH	144072
E. rufescens	Sudan	34	73	107	32%	FMNH	144105
E. rufescens	Tanzania	43	81	124	35%	FMNH	193293
E. rufescens	Sudan	37	76	113	33%	FMNH	144089
E. rufescens	Sudan	35	80	115	30%	FMNH	144106
E. rufescens	Tanzania	32	71	103	31%	RMCA	10833
E. rufescens	Sudan	34	71	105	32%	FMNH	144100
E. fuscipes	Zaire	34	49	83	41%	RMCA	1312
E. brachyrhynchus	Tanzania	29	60	89	32%	RMCA	8903
E. brachyrhynchus	Tanzania	31	59	90	34%	RMCA	12354
E. intufi	captive	35	60	95	37%	ZFMK	79496
Rhynchocyon:							
R. petersi	Tanzania	60	141	201	30%	FMNH	182515
R. petersi	Tanzania	77	120	197	39%	FMNH	161395
R. petersi	Tanzania	71	136	207	34%	FMNH	158851
R. petersi	Tanzania	67	110	177	38%	FMNH	163866
R. cirnei	Tanzania	71	130	201	35%	FMNH	171617
R. cirnei	Kisangani, Zaire	53	118	171	31%	RMCA	16508
Macroscelides:							
M. proboscideus	captive	31	55	86	36%	Zoo Halle	204067
M. proboscideus	captive	31	58	89	35%	ZooWupp	20602Wr
M. proboscideus	captive	30	56	86	35%	ZooBernb	860
M. proboscideus	captive	33	53	86	38%	ZooWupp	20202A
M. proboscideus	captive	34	52	86	40%	ZooWupp	20004J
M. proboscideus	captive	33	47	80	41%	ZooWupp	204004J
M. proboscideus	captive	29	56	85	34.%	ZooWupp	2040048
M. proboscideus	captive	30	53	83	36%	ZooHalle	203299
M. proboscideus	captive	11	28	39	28%	ZooWupp	206002R
M. proboscideus	captive	13	29	41	31%	ZooWupp	206002S
M. proboscideus	captive	12	33	45	27%	ZooWupp	207021
M. proboscideus	captive	22	43	65	33%	ZooHalle	205051r
M. proboscideus	captive	19	35	54	35%	ZooErfurt	no number
M. proboscideus	captive	13	27	40	32%	ZooWupp	20702L
M. proboscideus	captive	8	25	33	24%	ZooWupp	Fetus

Table 2.2 Measurements (in mm, except for penis location which is presented as %) for the distance between the anus and penis, penis and tip of lower lip, and anus to tip of lower lip for four different genera of Macroscelidea (presented as mean ± standard deviation and (range)).

Genus	Anus-penis	Penis – tip lower lip	Anus – tip lower lip	penis location (%)
Elephantulus (n = 16)	33.5 ± 3.6 (28 – 43)	66.5 ± 10.0 (49 – 81)	100.0 ± 11.6 (83 – 124)	33.7 ± 3.7 (29 – 41)
Macroscelides (n = 8)	31.4 ± 1.8 (29 – 34)	53.8 ± 3.4 (47 – 58)	85.1 ± 2.6 (80 – 89)	36.9 ± 2.5 (34 – 41)
Petrodromus (n = 7)	45.3 ± 6.8 (36 – 55)	109.6 ± 22.3 (83 – 137)	152.4 ± 22.6 (119 – 185)	29.9 ± 4.4 (24 – 36)
Rhynchocyon (n = 6)	66.5 ± 8.7 (53 – 77)	125.8 ± 11.8 (110 – 141)	192.3 ± 14.7 (171 – 207)	34.6 ± 3.7 (30 – 39)

All subsequent analyses were conducted with adult animals only. In *Elephantulus* (n= 16, represented by four species: *E. brachyrhynchus, E. fuscipes, E. intufi, E. rufescens*) the penis position varied from 29 to 41%, and *Rhynchocyon* (n= 6, represented by two species: *R. cirnei* and *R. petersi*) ranged from 30 to 39%. While the Shapiro-Wilk W tests for all groups rendered W values that could not reject the null hypothesis of normal distribution, the result for Petrodromus close (W = 0.835, P = 0.051). The data were then subjected to a one-way ANOVA, which indicated there was significant variation in the position of the penis among the four genera examined (F = 4.77, P < 0.01). A post-hoc Sheffe´-test indicated that the only comparison between individual genera that rendered significant differences was between *Petrodromus* and *Macroscelides* (p < 0.05). Figure 2.3 graphically illustrates the overlap among all four genera, at least in range.

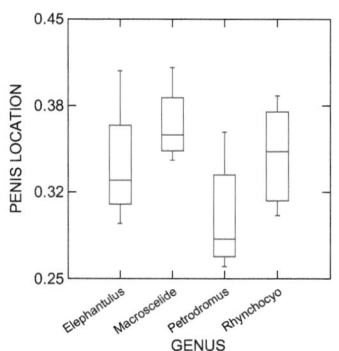

Even within the same species, there is much variation in penis position. For example in adult *M. proboscideus* the values range between 34 and 41%, and *P. tetradactylus* ranged from 24 to 36% (Table 3.1)

Fig. 2.3 The Box and Whisker Plot illustrates the overlap in ranges among all four genera.

2.3.2 Hyrax accounts

The mean for body length (BL) in *H. brucei* was 454mm (n = 116 in three studies of Barry and Shoshani 2000) and the distance penis-anus (PA) was 64.5 mm (n = 9, Coetzee 1966). On the bases of these measurements a percentage of 14.2 % for relative penis position was computed. The mean BL for *P. capensis* was 478 mm (n > 16 in two studies of Olds and Shoshani 1982) with a PA of 22.3 mm (n = 10, Coetzee 1966) which resulted in a percentage of 4.7 %.

2.4 Discussion

2.4.1 Distance anus-penis

Unlike the penial morphology (Stoch 1954; Woodall 1995a, b) and the ultrastructure of spermatozoa (Woodall and FitzGibbon 1995), the position of the penis can only distinguish two sengi genera: *Petrodromus* and *Macroscelides*. The significant difference in the location of the penis between young and adult *M. proboscideus* may be due to the changing body proportions during various growth stages. As data are available only for very young *M. proboscideus* and some of the slightly older age classes are lacking, it cannot be determined at which age the body proportions and penis positions reach the adult state. However, adult weight is attained at 46 days after birth (Rosenthal 1975), and normally at three months the animal attains sexual maturity (Olbricht et al. 2005) although in *E. rufescens* sexual maturity was achieved at 50 days post partum (Rathbun et al. 1981). The distance between anus and penis of 4 cm given in Sauer and Sauer (1971) for adult *M. proboscideus* was slightly greater than the distance of 3.4 cm found in this study.

The significant difference in the location of the penis between *Macroscelides* and *Petrodromus* corresponds to the distinctive osteological structures, such as proportional size of the bullae and the relative bizygomatic width (Evans 1942), and other morphological characters (Corbet and Hanks 1968). While *Petrodromus* shares some features of penis morphology with other members of the Macroscelidinae, *Macroscelides* is exceptional in this respect (Woodall 1995b). Considering the penis shape a marker of evolutionary history, Douady et al. (2003) used molecular methods and found significant differences between *Macroscelides* and the *Petrodromus/E.rozeti* clade. This is in contrast to allozyme and isozyme analyses where the degree of divergence between *Macroscelides* and *Petrodromus* was low (Raman and Perrin 1997). Given the intriguingly close relationship between *Petrodromus* and *E. rozeti* (Douady et al. 2003), examination of the external genitalia of male *E. rozeti* would be particularly interesting.

2.4.2 Genital morphology in the Afrotheria

Afrotherian mammals exhibit some unusual genital features including penis location and morphology. The reproductive system of male sengis show similarities with that of paengulates (hyraxes, elephants and sirenians, Woodall 1995a, Woodall and Skinner 1995).This study revealed that the genital opening in male sengis is located very close to the umbilical scar which is similar to male sirenians (Reynolds and Odell 1991). But although the penis of both sengis and tenrecs is especially long, the latter is located in a false cloaca. Desert golden moles (*Eremitalpa granti*) evolved a unique feature among the Afrotheria with the penis being situated within the cloaca (Perrin and Fielden 1999). The presence of these unusual penial features provokes the question why they have evolved and how these anatomical constraints influence the behaviour of the various afrotheres.

2.4.3 Copulation posture

Afrotherian mammals exhibit a longer vertebral column than other mammals (Sánchez-Villagra et al. 2007). Sengis possess a long, although shorter than paengulates, but relatively rigid lumbar region resulting from a beamlike reinforcement in two sections of the vertebral column which prevents sagittal bending. This structure is related to their mode of locomotion (Frey 1994a). Due to this lack of sagittal bending, which implies reduced mobility (Frey 1994a) the approach of the male to the female genital region is difficult. Nevertheless, Sauer (1973) observed male *M. proboscideus* mounting dextrously despite their rotund body shape, the long thin penis rapidly finds the vagina. Lumpkin and Koontz (1986) found male *E. rufescens* standing almost erect on the hind limbs while the female had adopted a lordotic posture in which her hind legs were almost fully extended and her rump was elevated. Dynamic sagittal flexions are almost impossible. In sengis this is balanced by the standing copulation posture which implies a cranio-ventral penis position, whereas sirenians with their completely rigid rump copulate facing one another while swimming horizontally (Reynolds and Odell 1991). Elephants have also a relatively rigid rump which is compensated for copulation by particular genital structures in both the male and the female (Frey 1994a).

Copulation in *M. proboscideus* took less than 10 seconds (Rosenthal 1975). The pars libera penis, the part of the penis which is not attached to the ventral semilunar line, is short in sengis and allows intromission only of the most anterior part of the penis tip (Frey 1994a). This short copulation time may be part of risk-sensitive behaviour. In the Tenrecinae the long penis penetrates the vagina up to the base and the baculum most probably is useful here. The mount lasts much longer than in sengis (Eisenberg and Gould 1970).

2.4.4 Hyracoidea

The elongate vertebral column in hyraxes is longer but less rigid than in other afrotheres and enables more parasagittal flexibility (Sánchez-Villagra et al. 2007). Therefore, they do not need a long penis. In *Procavia,* penis length is only 8 % in comparison to sengis with up to 75% of the body length. The relative position of the penis of all sengi genera was far more anteriorally situated than in the two genera of the Hyracoidea which were examined here. In *P. capensis* the penis is situated closer to the anus with a distance anus–penis of 4.7% of BL in comparison to *H. brucei* with 14.2%. In sengis the relative penis position ranged from 29 to 41%.

However, the percentage cannot be compared directly with the percentage computed for the sengis since the accounts for body length (BL) were not taken in the same way and thus do not correspond to the total length (TL) taken for the sengis.

2.5 Conclusions

The results suggest the position of the penis is of utility in differentiating between two genera of the Macroscelidae: *Petrodromus* and *Macroscelides*. Comparisons between any other sets of sengi genera revealed broad overlap in penis position. Comparisons between adult and very young *Macroscelides* revealed significant differences in penis position between the two age groups. Further sampling is desirable to refine our understanding the change of penis position as sengis age. Penis location might then serve as one of many other features to determine the time of puberty. Additional samples would also further test our conclusions and place them in the context of other taxonomic studies that have utilized other morphological features and molecular data.

The penis of sengis is located far cranially which is a consequence of other anatomical constraints such as a rigid lumbar region, penis length, and internal testes. They share some features of the male reproductive system with other members of the Afrotheria.

CHAPTER 3

3. THE TAXONOMIC DISTRIBUTION OF TEATS IN SENGIS

(Data were partially obtained in collaboration with Dr. T. William Stanley, Field Museum of Natural History, 1400 South Lake Shore Drive, Chicago, IL 60605, USA.)

3.1 Abstract

The female reproductive organs of sengis are well studied but information on exact teat location is scarce and contradictory. Males and females of all four sengi genera were examined for teats. Only in two species teats were found on males: *Elephantulus rozeti* and *Petrodromus tetradactylus* but not on other species examined. This supports the recent taxonomic conclusions regarding the relationship of these two taxa. However, teat position is not useful in taxonomic distinction among the four genera.

Altogether 25 female and 2 male specimens were measured and analyzed in order to determine exact teat locations along the longitudinal axis as well as in relation to the body circumference. The arrangement of teats was determined in a formula for each genus which distinguishes between antebrachial, abdominal and inguinal regions. No sengi exhibited teats situated dorsolaterally to the extent of those in rock-dwelling mammals such as *Petromus typicus*.

3.2 Introduction

Especially the molecular work on placental phylogeny has provided strong support for the monophyly of the Afrotheria as a supraordinal clade of living African mammals (Springer et al. 1997, Stanhope et al. 1998a, b) but morphological information is accumulating slowly (Asher and Lehman 2008, Tabuce et al. 2008).

While both historical and recent studies have focused on the natural history of this enigmatic group, limited or conflicting information is available on the topography of the reproductive features such as the teats of the Macroscelidea. The number or placement of teats in females have long been utilized to distinguish among different species or genera of mammals (Corbet and Hanks 1968, Kunz et al. 1996) or generate hypotheses on the life histories of various mammals with unusual number or placement of teats such as *Mastomys* or *Myocastor* (Martin et al. 2001).

Numerous publications are available with detailed descriptions of the female reproductive organs of sengis (Macroscelidea) (e.g. van der Horst 1944, e.g. van der Horst and Gillman 1941a, b, Tripp 1970, McKerrow 1994). Although there have been morphological examinations of different sengi species including recording the number and location of mammae in females (Fitzinger 1867, Shortridge 1934, Corbet and Hanks 1968, Smithers and Skinner 1990, Skinner and Chimimba 2005), this information is often ambiguous since various terms were used by different authors for the same teat location.

3.2.1 Nomenclature of teat positions

Terms such as axillary, antebrachial, pectoral, thoracic, abdominal, inguinal and pubic are in use with no additional information on the precise boundaries of these regions (e.g. Smithers 1983, Simons 1993, Kunz et al. 1996). Mostly the number of teat pairs in each anatomical region such as pectoral (P), abdominal (A) and inguinal (I) was for example expressed for an animal with three teat pairs as P1, A1 and I1. The term antebrachial was used for *P. tetradactylus* (Jennings and Rathbun 2001) which means in front of the forelimb, and "nuchal" teats are published for sengis by Fitzsimons (1942) and Corbet and Hanks (1968). Although Smithers and Skinner (1990) provided an illustration for the general location of teats in female mammals and how to term those parts, for most sengi species there is a lack of illustrations or pictures which clearly show and term the exact locations of the teats. Sauer and Sauer (1972) observed a nursing female *Macroscelides proboscideus* in the wild but these descriptions together with illustrations and pictures of the nursing young (Rosenthal 1975, Vakhrusheva 2000 – *M. proboscideus*, Rathbun 1979a, b - *Rhynchocyon chrysopygus*) can give only a vague idea. The only illustration for *Elephantulus myurus jamesoni* is incomplete (McKerrow 1954).

Some authors make a distinction between the terms "teat" and "teat" which depends on the number of galactophores present (Blüm 1985) or the presence or absence of a common reservoir from which milk is expressed through a single duct (Martin et al. 2001). For the purpose of this study I refer to the external structure as mammae or teats. Until this study, the presence of teats in male sengis has never been mentioned to my knowledge.

3.2.2 Behavioural ecology

Sengis occupy various ecological niches including dry habitats such as the Namib Desert (*M. proboscideus*) to humid tropical regions such as the coastal forests of eastern Africa (*R. chrysopygus*). Such differences in habitat are presumably reflected in a high variability in adaptations within the order. The location of the mammae has been suggested to be adaptive and varies in accordance with specific nursing requirements (Long 1969). Such a hypothesis depends on a detailed description of the existence and location of teats on each sex of the different sengi species.

3.2.3 Aim of the study

In this study I examine the topographic position of teats using specimens from all four genera of the order, to determine whether 1) mammae are present in both genders, 2) there are interspecific or intergeneric differences in teat location within some genera in the order Macroscelidea, and 3) to determine whether some sengis exhibit mammae located as high up dorsolaterally as those found in rock-dwelling mammals such as *Petromus*.

3.3 Material and Methods

3.3.1 Material

Altogether 26 specimens from eight different sengi species representing all four genera within the order, and from seven different locations were examined and measured (Table 3.1). Some of the specimens were wild caught, and others were born in captivity. The age of captive animals was recorded, when known. All specimens were preserved in 70% ethanol or 5% formalin.

3.3.2 Methods

Measurements were taken by myself and the three colleagues, Dr. William Stanley, Lucas Thibedi, and Beryl Wilson, using calliper and ruler. For each teat pair on one side of the longitudinal axis of the body, measurements were taken to determine the distance of each teat pair to the lower lip (NL) and anus (NA) (Fig. 3.1). Each teat pair was numbered from anterior to posterior, i.e. teat pair 1 is the most anterior pair, so the measurements were labelled NL^1 and NA^1. NL and NA were summed for each teat, and the relative position was expressed as a percentage of NA/SUM. Both

Petrodromus and *Rhynchocyon* had only two pairs of teats. The former lacked a pair in the posterior third of the abdominal region, so the mammae were identified as N^1 and N^2. *Rhynchocyon* lacked mammae in the anterior third of the venter, so teats were labelled N^2 and N^3.

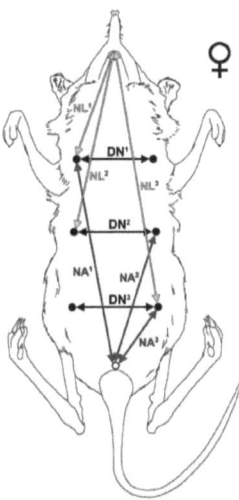

Fig. 3.1 Measurement methodology to determine teat locations (skematic drawing by R. Banasiak).

Subsequently, the location of the teats along the plane perpendicular to the longitudinal axis was determined. For this measurement, a string was run completely around the circumference of the body at the teat along the plane perpendicular to the axis of the body, and the length of the amount of string to encompass the body was measured (CN for circumference at teat). The distance between the teats of the same pair was marked on the string and measured (DN for distance between teats). The percentage resulting from DN/CN was compared among species.

This method was also used to investigate if, and to what extent "lateral" or "dorsal" teats exist (i.e. those teats found in *Petromus*). The greater the percentage for a given teat, the more "lateral" it was

While the quality of the material differed due to preservation methods as well as subsequent treatment (such as removal of the skull or ventral incisions), the use of ratios involving teat location relative to either anus and lower lip, or circumference of the body allowed suitable comparison among different specimens. Standard descriptive statistics (mean, range, and standard deviation) were calculated for measurements for each feature.

3.4 Results

3.4.1 Locations of the teats

3.4.1.1 Males

There was no evidence for teats in males of any of the sengis except *Petrodromus tetradactylus* and *Elephantulus rozeti* (Fig. 3.2) however, only one individual of each species was measured. In each, the teat location corresponded generally to the placement observed in females of the same species, but the male teats were much smaller than those of the females. Teats in a male *Petrodromus* were measured with 1.52 mm diameter on the bases of the antebrachial teate in contrast to 2.6 mm (antebrachial) and 1.56 mm (inguinal) in a female.

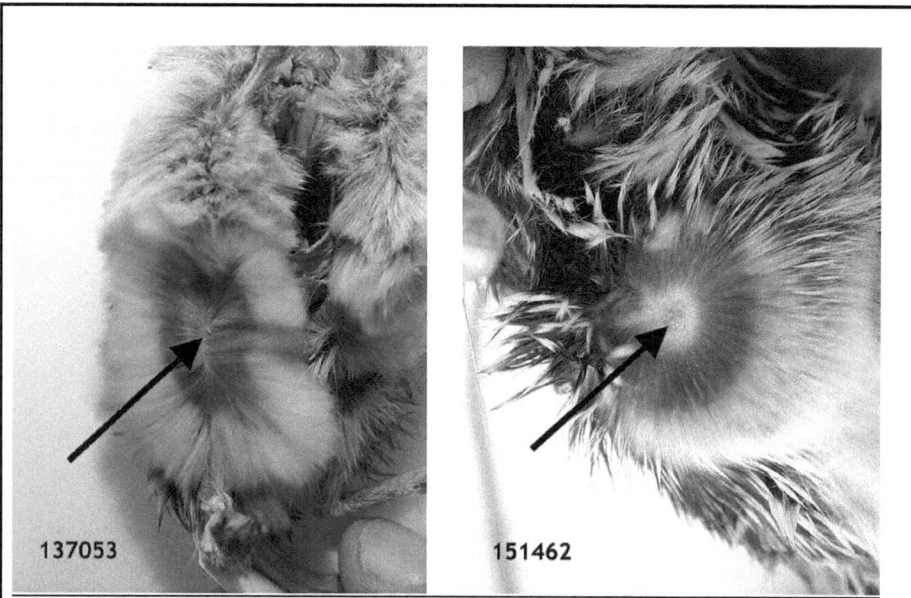

Fig 3.2 Male teats in *Elephantulus rozeti* (left, FMNH 137053) and *Petrodromus tetradactylus* (right, FMNH 151462), photos by R. Banasiak

3.4.1.2 *Females*

In view of teat location and number on females of the four sengi genera, the differences are striking. *Rhynchocyon* and *Petrodromus* both have only two teat pairs but there are variances in teat position (Fig 3.3).

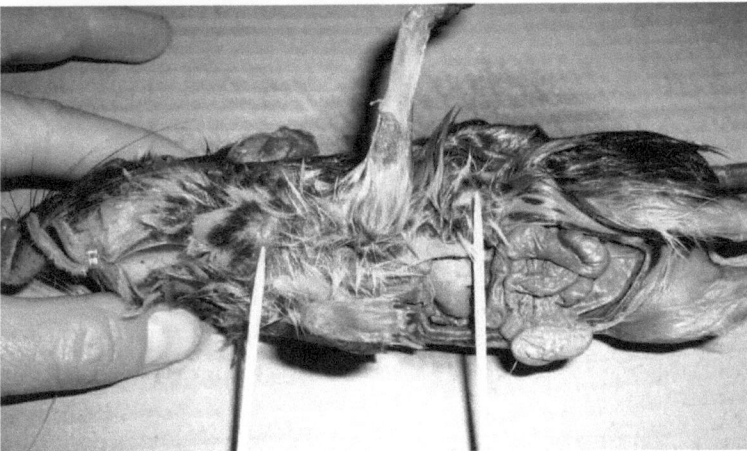

Fig 3.3 Two teat pairs in *Rhynchocyon chrysogygus* (top, photo by G. Rathbun) and *Petrodromus tetradactylus* (bottom)

In contrast, female *Macroscelides* and *Elephantulus* have 3 teat pairs in the same position (Fig. 3.4.).

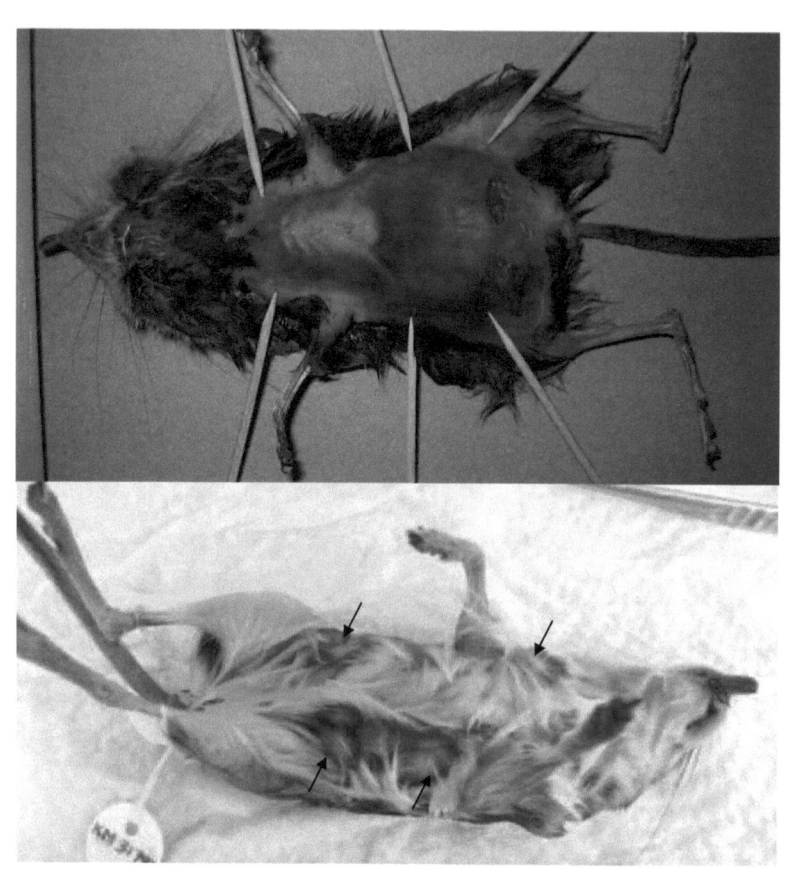

Fig. 3.4 Three teat pairs in female *Macroscelides proboscideus* (top) and *Elephantulus rupestris* (bottom, photo by L. Thibedi, Amathole Museum). Arrows denote 4 of the six teats.

The relative longitudinal positions of teats in females of the four genera examined, including six *Elephantulus* species (*E. brachyrhynchus*, E. edwardii, *E. intufi, E. rozeti, E. rufescens, E. rupestris*), *M. proboscideus, P. tetradactylus* and *Rh. cirnei* are given in Table 3.1 and the means are listed in Table 3.2.

DISTRIBUTION OF TEATS

Table 3.1. Measurements of teat placement in female sengis. Teats are numbered from anterior to posterior (i.e. 3 is closest to anus); S = Source (W-wild born, C-captive); NL = distance between teat and anterior tip of lower lip; NA = distance between teat and anus; PN = relative position of teats along the longitudinal axis of the body; CN = circumference of body at plane of teat; DN = distance between the two teats of a pair; DN/CN = relative lateral position of teats

Museum & #	Species	S	NL¹	NL²	NL³	NA¹	NA²	NA³	PN¹	PN²	PN³	CN¹	CN²	CN³	DN¹	DN²	DN³	DN/CN¹	DN/CN²	DN/CN³
FMNH 47749	*Elephantulus brachyrhynchus*	W	28	48	60	63	40	29	31%	55%	67%		97	104		28	25	29%	29%	24%
RMCA 11704	*E. brachyrhynchus*	W	31	51	67	65	41	27	32%	55%	71%									
RMCA 11185	*E. brachyrhynchus*	W	31	51	68	64	40	28	33%	56%	71%									
McGreg	*E. edwardii*	W	29	51	74	69	51	32	30%	50%	70%		96			31			32%	
ZFMK 79493	*E. intufi*	W	28	48	61			30	30%		67%									
ZFMK 79497	*E. intufi*	C	25	47		51	32		33%	59%										
ZFMK 79499	*E. intufi*	C	28	44	56	52	35	25	35%	56%	69%									
FMNH 137052	*E. rozeti*	C	35	56	66	57	35	26	38%	62%	72%	80	83		22			28%		
RMCA 7449	*E. rufescens*	W	43	65	86	75	48	30	36%	58%	74%									
Amathole KM31788	*E. rupestris*	W										78	99	96	17	23	19	22%	23%	20%
Amathole KM31790	*E. rupestris*	W										84	102	97	22	30	21	27%	29%	22%
Amathole KM31791	*E. rupestris*	W										81	108	107	19	31	23	24%	29%	21%
Amathole KM31795	*E. rupestris*	W										72	87	88	16	26	20	22%	30%	23%
Amathole KM31798	*E. rupestris*	W										73	99	92	19	29	19	26%	29%	21%
ZooWupp 20301L	*Macroscelides proboscideus*	C					40													
ZooWupp 99021F	*M. proboscideus*	C	31	54	69	58	33	26	35%	62%	73%	88	82	104	21	28	28	24%	34%	27%
ZooWupp 20112F	*M. proboscideus*	C	27	49	60	65	42	33	29%	54%	65%	82	81	92	18	28	27	22%	34%	29%
ZooWupp 205071	*M. proboscideus*	C	29	52	67	58	43	29	33%	55%	70%	90	93	100	21	29				
ZooWupp 20202C	*M. proboscideus*	C	36	62	75	56	35	26	39%	64%	74%									
ZooWupp 2040047	*M. proboscideus*	C	31	50	65	58	41	23	35%	55%	74%	93	85	102	17	25	20	18%	30%	20%
ZooHalle203230	*M. proboscideus*	C	27	53		56	36	28	33%	60%										
FMNH 151459	*Petrodromus tetradactylus*	W				110	66													
FMNH 169664	*P. tetradactylus*	W	48	83		89	51		35%	62%		100	117		14	17		14%	15%	
RMCA97009M5207	*P. tetradactylus*	W	55	92		104	72		35%	56%										
RMCA A1097M404	*Rhynchocyon cirnei*	W		149	187		76	47		66%	80%									
HumbUni Berlin 58	*Petromus typicus*	C		96	122		77	53					128	121		67	72		52	59

47

The relative position of teat pair 1 (the most anterior) ranged from 29 to 39% (both values recorded in *M. proboscideus*), with the means for the three genera that had mammae in that general position (i.e. not including *Rhynchocyon*) being 34% for *Elephantulus* (n = 8) and *Macroscelides* (n = 6) and 35% for *Petrodromus* (n = 2). The range of teat pair 2 ranged from 50% (*E. edwardii*) to 66% (*Rhynchocyon*), and the mean value was 56% (*Elephantulus*; n = 8), 58% (*Macroscelides*; n = 6), 59% (*Petrodromus*; n = 2) and 66% (*Rhynchocyon*; n = 1). Teat pair 3 ranged from 65% (*Macroscelides*) to 80% (*Rhynchocyon*), and the mean for the genera that had teats in that region (i.e. not including *Petrodromus*) was 70% (*Elephantulus*, n = 8), 71% (*Macroscelides*; n = 6) and 83% (*Rhynchocyon*; n = 1).

Table 3.2 Horizontal and vertical location of teat pairs. The vertical value is the percentage of the location of a teat pair in relation to the total body length (distance lip-anus, see Fig, 3.1 with DN^1, NL^1, NA^1 for pair 1). The horizontal value is the percentage of the distance between teats in a pair in relation to the circumference where the teat pair is located. N.P. means not present. Values given as mean ± standard deviation, (range) and sample size.

Genus	Vertical (%)			Horizontal (%)		
	Teat pair 1	Teat pair 2	Teat pair 3	Teat pair 1	Teat pair 2	Teat pair 3
Macroscelides	34 ± 3.3 (29-39) n = 6	58 ± 4.2 (54-64) n = 6	71 ± 3.8 (65-74) n = 5	21 ± 3.1 (18-24) n = 3	33 ± 2.3 (30-34) n = 3	25 ± 4.7 (20-29) n = 3
Elephantulus	34 ± 2.7 (30-38) n = 8	56 ± 3.5 (50-62) n = 8	70 ± 2.4 (67-74) n = 8	25 ± 2.6 (22-28) n = 6	29 ± 2.7 (23-32) n = 7	22 ± 1.5 (20-24) n = 6
Petrodromus	35 ± 0.0 (35-35) n = 2	59 ± 4.2 (56-62) n = 2		14 n = 1	15 n = 1	
Rhyncnhocyon		66 n = 1	83 n = 1			
Petromus		55 n = 1	70 n = 1		52 n = 1	59 n = 1

Greater variation exists for the horizontal position of the teat pairs among genera (Table 3.2). However, none of the sengi genera examined exhibited a value as large as that for *Petromus typicus* (52%; Fig. 3.5). The largest value for sengis was that of *M. proboscideus* (34%), and the smallest was *P. tetradactylus* (14%). *Rhynchocyon* was not measured, but teats are clearly contained on the venter (Fig. 3.3).

Whether the term lateral or dorsal is applied to the most dorsal teat pair in *Petromus*, it is clear that there is no corresponding level of laterally elevated teats in sengis.

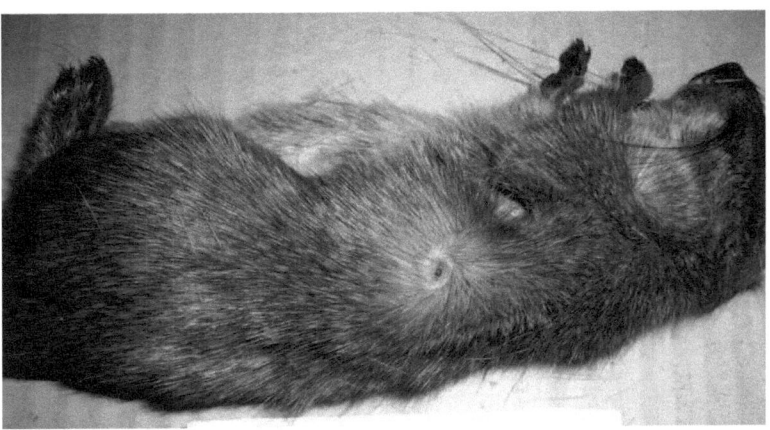

Fig. 3.5 Dorsolateral teat in a female rodent, *Petromus typicus*

3.4.1.3 Formulas

Teat formulas refer to the longitudinal axis which meets the values for the vertical teat pairs. Definitions for pectoral (over the rib cage), abdominal (below the rib cage, anterior to the pelvic girdle), and inguinal (in the angle of the hind limbs, close to the pelvic bones) are the base for formulas of teat locations. Since applying these definitions may sometimes be difficult due to the stiff fixation of the specimen the percentages of the distance of a particular teat pairs to the lip in relation to the body length are the basis of our results. Based on the means for teat positions of the two or three teat pairs a rough arrangement could be defined for all four sengi genera as follows: the first teat pair is located at about one third of the body length starting from the lip (34 and 35%), the second pair a little less than two thirds (56, 58 and 59%) and the third pair at about 3 quarters (70, 71 and 83%).

The most anterior pair is located above the rib cage very high up closed to the clavicula. So the first number stands for the antebrachial position (teat pair 1), followed by abdominal (teat pair 2) and inguinal (teat pair 3). The following formulas are proposed:

Rhynchocyon: 0-2-2
Petrodromus: 2-2-0
Macroscelides: and *Elephantulus*:   2-2-2

3.5 Discussion

It is clear that there is the potential for a number of measurement errors. Firstly, the quality of the material differed enormously. Standardizing measurements such as using always the same body side was not possible due to damage or distortion of the body. Measuring soft tissue bears the general problem that tissue can be stretched out too much while working on the body. This applies in particular with specimens where the skull was removed. This problem was often combined with a ventral cut thru the lip so that an accurate measuring spot was difficult to detect without stretching the upper part of the body into an abnormal position. Alternatively, the tissue can be stiff and shrunk due to various fixation methods.

Steiner & Raczynski (1976) stress in their approach to identify measurement errors that a repetition of the measurement is worth only where a single or only a few specimens are of special importance. Therefore, some of the extreme accounts were double checked.

3.5.1 Teats on male sengis

Mammary glands in males mostly are miniature versions of those of the female (Cooper 1840) and these results for male *Petrodromus* and *E. rozeti* are in line with this observation. Males of different sengi species have never been observed taking part in the care of young (Sauer 1973, Rathbun 1979a, Baker et al. 2005) and therefore, lactating males are not expected. If present, teats or teats in male mammals usually are not only small but not functional (Blüm 1985). The males of many rodents are exceptional since they are lacking teats (Mares 1988, Imperato-McGinley et al. 1986, Gilbert 2006) and male horses are not only lacking teats but most often also mammary tissue (Ellenberger and Baum 1974). If mammary glands in males are active, this is mostly considered as an abnormality. Such polymastia is known for primates (Hartman 1927) and humans (Cooper 1840). Functional teats as a normal pattern are described for the Malaysian fruit bat (Francis et al. 1994) where the males regularly feed their young. Notably, some male bats lack abdominal (pubic) teats in contrast to females (Simmons 1993). Further research will elucidate potential functionality of the mammae which were observed in male sengis (Olbricht and Welsch in prep.). More detailed information is available in chapter 4.

That teats were found in males of only *Petrodromus* and *E. rozeti* is intriguing given recent taxonomic phylogenetic studies involving these taxa. Woodall (1995a, b) found similar structures in male reproductive organs between *E. rozeti* and *Petrodromus*. His morphological findings were subjected to molecular research which confirmed a close relationship between the two species

(Douady et al 2003). The authors suggested that the basic morphology was maintained over the course of about 11 million years since the separation of *Petrodromus* and *E. rozeti*. They also presume that the aridification of the Sahara might be an explanation for the similarity in reproductive structures of these two species since in such a vicariant event there would not be strong selection for differentiation of reproductive structures. A more detailed study of the sexual structures of both *Petrodromus* and *E. rozeti* may further elucidate the relationship between these two taxa.

It can only be speculated whether the existence of male teats in the two sengi species is a relict in the sense of phylogenetic inertia or whether it points to the evolutionary future of these animals. The close relationship between *Petrodromus* and *E. rozeti* supports the hypothesis of phylogenetic inertia. On the other hand, although a very ancient mammalian group, sengis generally show a high potential to adaptation, such as omnivorous diet (Kerley 1995), physiological traits e.g. torpor (Lovegrove et al. 2001a, b), and reproductive traits (Sauer 1972) and have populated a number of different habitats.

3.5.2 Teats on female sengis and formulas

Teat positions with precise boundaries are identified for all sengi genera and formulas were developed on this basis. The following modifications were made although the relative position of teats was not useful in taxonomic distinctions between genera.

Corbet and Hanks (1968) used the teat location as a diagnostic character for the distinction of the two subfamilies of Macroscelididae. Their observations reveal two pairs of abdominal mammae in the Rhynchocyoninae whereas *Petrodromus* lacks abdominal mammae. Smithers and Skinner (1990) had stated two pairs of pectoral teats for *Petrodromus*, Corbet and Hanks (1968) one antebrachial and one pectoral pair. In contrast to both, in this study one pectoral and one abdominal teat pair were found with the inguinal pair missing. For *Rhynchocyon* one abdominal and one inguinal pair were determined in this study.

Terms for teat locations in *Macroscelides* and *Elephantulus* were modified from antebrachial, pectoral and abdominal (Corbet and Hanks 1968) to antebrachial/pectoral, abdominal and inguinal. Mc Kerrow (1954) had erroneously found only two pairs of teats instead of three in *Elephantulus myurus jamesoni* and determined their positions as "thoracic and abdominal pair". In fact, her illustration showed more an antebrachial/pectoral and inguinal position of two teat pairs which is indeed more realistic with only the central pair missing. But she concludes in light of the large expanse of mammary tissue, that "during lactation the glands extend to the inguinal region of the body". Her observation most probably refers to the third teat pair which was not found in that study.

Furthermore, Corbet and Hanks (1968) declare the presence of nuchal, pectoral and abdominal teats for the two genera *Macroscelides* and *Elephantulus* as distinctive characters to the other two sengi genera. The term nuchal (see section below) was also used by Fitzsimons (1920). In both publications the term "nuchal" was later replaced by the term "antebrachial".

3.5.3 The enigma about nuchal, lateral and dorsal teats

3.5.3.1 Historical observations

Fitzsimons (1920) published the observations of Bertie van Musschenbrock who observed young rock jumping shrews (meaning sengis) "hanging on to something on top of the shoulder blades and a later examination revealed two teats, one on each side". In these early years he determined this species "inhabiting the Cape Province both east and west" as rock jumping shrew or rock elephant-shrew (*Elephantulus rupestris typicus*) but because of the confusing taxonomy of that era, there are other possibilities such as *E. edwardii* and *E. myurus* (Skinner and Chimimba 2005). Another unusual account was documented by Hufnagl (1972) in a citation of Desmond Morris who observed young *E. rozeti* which were "carried around by her mother, attached to her teats." It is doubtful that the young were riding on their mother´s back being attached to a dorsal teat since teat clinging cannot be expected in this highly precocial *Elephantulus*-species. This behaviour was never documented on another occasion. Regardless of the species, the published observation of "lateral" or "dorsal" teats in a sengi is intriguing. The results of this study do not confirm the existence of nuchal or dorsal teats.

3.5.3.2 The presence of dorsolateral teats

Dorsal or lateral teats are found in mammal species such as those within the rodent genera *Petromus*, *Hystrix* and *Myocaster*. This teat location might allow a particular nursing position that is adaptive in some habitat settings such as narrow spaces where i.e. a rock dweller lives. However, the exact terms and boundaries of these regions are ambiguous and observations sometimes are not reliable. The nutria (*Myocastor coypus*) rests in shallow water in reed nests (Puschmann 2004). Lying flat on the riverside might be comparable to a rock dweller when lying flat under rock crevices. The eight to twelve teats are so far laterated that they are conspicuous when the female is viewed from above and giving rise to the erroneous statement that they are even located in the back (Lowery 1974). Another example is the crested porcupine (*Hystrix cristata*) which sleeps in rock crevices (Grubb et al. 1998). It has "two lateral pairs of mammae right after the shoulders" (Mohr 1965). The dassie-rat or noki (*Petromus typicus*) with its strictly rock dwelling life style has two lateral and one dorsal teat pair. Habitat setting or life style of these rodents are comparable to some sengis such as the North African sengi *(E. rozeti)*, the Western Rock sengi *(E. rupestris)*, the Eastern Rock sengi *(E. myurus)* and the Cape sengi *(E. edwardii)*. As a consequence, similar teat locations of all rock-dwellers are to be expected as an adaptation to living in rock crevices and to the absence of nests in some species but the results of this study do not correspond to these speculations.

3.5.3.3 Defining a dorsolateral teat location

Apart from anecdotal observations attempts have been made by scientist to determine a lateral position for mammae. Emmons and Feer (1990) illustrated teat locations for the Echimyidae but defined this region as covered by dorsal fur which is a very vague description. Using the methodology of this study for determining the lateral position of mammae, a female noki was measured (Fig. 3.5) and these data were compared with sengis. None of the sengis which were examined exhibited teats as far up laterally as *Petromus* (59%, Tables 3.1 and 3.2). Thus, I hypothesize that the observations of Van Musschenbrocks (as stated in Fitzsimons 1920) might have been a mis-interpretation of a relatively rare behaviour of pedal scent-marking, documented in *E. rufescens* (Rathbun and Redford 1981). I also speculate that rock-dwelling rodents such as *Petromus* are more strictly adapted to their habitat and life style than petrophile sengis and thus, their morphological traits such as teat location are more specialized. The results of this study suggest that dorsolateral teats are not present in sengis.

3.5.3.4 Functionality of teats

Some authors have noted the non-functionality of some teats in other mammals, e.g the anterior teat pair of small diprotodont marsupials (Ward 1998). The noki (*Petromus typicus*) has "two pairs of mammae situated on the sides of the body just behind the shoulders and a third pair further back on the body" (Skinner and Smithers 1990) which often is non-functional (De Graaf 1981) or individually not present (Chris and Tilde Stuart pers. comm., Andrea Mess pers. comm.). In contrast, suckling was observed on teats of all regions in *Macroscelides* (Olbricht pers. observ) and *E. rufescens* (Rathbun pers. comm.).

3.6 Conclusions

The relative position of mammae was in taxonomic distinctions among the 3 of the four sengi genera: only between *Macroscelides* and *Elephantulus* there was no difference. The arrangement of teats could be determined in a formula for each genus. No sengi, including petrophile species, exhibited teats situated dorsolaterally to the extent of those in rock-dwelling mammals such as *Petromus typicus*

The results of this study support the findings of Douady et al. (2003) in finding the existence of teats in only male *Elephantulus rozeti* and *Petrodromus tetradactylus*.

CHAPTER 4

4. HISTOLOGICAL AND HISTOCHEMICAL STUDY OF THE MAMMARY GLANDS IN *PETRODROMUS TETRADACTYLUS*

(histological sections and stains were done in the Department of Anatomy, University of Munich)

4.1 Abstract

The mammary tissue of a female and a male wild-born *Petrodromus tetradactylus* was examined with different histological and histochemical methods. The results of this study reveal a potentially functioning mammary gland in male *Petrodromus* with evidence of active mammary tissue. The secretory units (acini) are sexually dimorphic. In females typical acini, milk ducts, cisternal milk sinus and a teat canal can be distinguished. The acini of the females occur in the periphery of the gland. Acini in the male teat occur in the connective tissue of the teat. These acini differ in morphology from those of the female, they are small and their lumen varies between narrow and comparatively wide. Additionally, in female and male *Petrodromus* scent glands are to be found at the base of the teat.

Only in *P. tetradactylus* and *E. rozeti* of the existing 16 sengi species, males have evidence of teats. Males of none of the sengi species contribute to the raising of their young and therefore, the function of potentially functional mammary tissue in male *Petrodromus* is not clear.

4.2 Introduction

Numerous publications are available with detailed descriptions of the reproductive organs and patterns of female sengis (Macroscelidea) (e.g. van der Horst 1944, e.g. van der Horst and Gillman 1941a and b, Tripp 1971, McKerrow 1954) but little attention has been paid to the mammary glands of sengis. The presence of teats in male sengis has never been mentioned in previous studies and was documented only recently by Olbricht and Stanley (subm.), who found that out of 16 sengis species males of only in two species, *Petrodromus tetradactylus* and *Elephantulus rozeti*, there is evidence for the presence of teats.

4.2.1 Mammae in male mammals

In general, mammary glands are functional only in the female, though homologues of the glands are present in the male. Natural selection has never favoured the production of milk by males (Eisenberg 1981) and teats in male mammals are usually small and non-functional (Blüm 1985). Rudimentary teats are found in almost all male mammals with the exception of some rodents (Imperato-McGinley et al. 1986, Gilbert 2006, Mares 1988), whereas male horses mostly are lacking mammary tissue (Ellenberger and Baum 1974). In primates, elephants and bats several anlagen (primordia) are formed but only one pair of mammary glands continues its development and remains evident in the adult male (Blüm 1985). Mammary anlagen were found in both sexes of the marsupials *Didelphis virginiana* and *Monodelphis domestica* but the neonatal males had less than one-third of the full female complement of mammary glands (Renfree et al. 1990).

On the other hand, according to Renfree et al. (1990), mammary glands can retain the ability to lactate if provided with the appropriate hormonal stimulus. When mammary glands in males become active, this is mostly considered as an abnormality. Such galactorrhoea is known from primates (Hartman 1927) including humans (Cooper 1840). Functional teats and mammary glandular tissue as a normal pattern are described for the Malaysian fruit bat (*Dyacopterus spadiceus*, Francis et al. 1994) where the males regularly feed their young.

4.2.2 Aim of the study

Teats of males are mostly miniature versions of the females of the same species (Cooper 1840), which was also true for male *Petrodromus* and *E. rozeti* (Olbricht and Stanley subm.). The purpose of this study was to investigate whether mammary tissue is found in the teat of male *Petrodromus* and to compare it at a histological and histo-chemical level with the glandular tissue of a female.

4.3 Material and Methods

4.3.1 Material

A female (RMCA A1.097-M-0404, ID from the Royal Muesum for Central Africa in Tervuren, Belgium or Munich 302-07, ID from the University of Munich) and male (RMCA A1.097-M-0403, Munich 216-07 and 217-07) *Petrodromus tetradactylus* originating from Djabir, Democratic Republic of Congo were used for histological examination. The specimens were wild born and a donation of the Royal Museum of Central Africa in Tervuren, Belgium. They were preserved in formalin. Tissue samples from the thoracic region were taken and subsequently imbedded in paraffin blocks for further examination under the microscope.

4.3.2 Methods

4.3.2.1 Fixation and embedding

Processing of the material for histology and histochemistry was done according to standard methods (Romeis 1989, Bancroft and Stevens 1990).

Because of the unknown composition of the initial fixative the specimens were refixed in 4.5% formaldehyde, and subsequently embedded in paraffin. Four µm thick paraffin sections were stained with the following methods. The detailed procedures are available in Appendix C.

- Haematoxylin and eosin (HE)

- Heidenhain's azan stain

- Alcian blue stain (pH 2.5)

- Periodic-Acid-Schiff technique (PAS)

- Immunohistochemical stain for actin

For the immunohistochemical demonstration of actin (Sigma, Deisenhofen) monoclonal antibodies of mice were used. After blocking the endogene peroxidase under exclusion of light by 3% H_2O_2 for five minutes the following staining steps were all undertaken with Histostain Plus Kit by Zymed (Carlton Court, San Franciscso, CA, USA). After rinsing with PBS (phosphate buffered saline), three drops of the second blocking reagent (with "normal" serum) are applied 15 minutes. Directly afterwards 100 µl each of the primary antibody are applied in a dilution of 1:400. It is washed off with PBS twice after 60 minutes time of influence at room temperature. Then 3 drops of biotinylated secondary antibody are applied and rinsed with PBS after 10 minutes. Afterwards streptavidin (conjugated with peroxidase) is put for 10 minutes on the sections and then rinsed twice with PBS again. At the end, 3.3-diaminobenzidine tetrahydrochloride plus 2 µl 30 % H_2O_2 is pipetted thereon for 10 minutes. After a rinse in Aqua dest., the sections eventually are dehydrated in ascending alcohols, cleared in xylol and mounted in DPX.

4.3.2.2 Histochemistry of lectins

This technique demonstrates specific carbohydrate components. The following lectins were used: *Helix pomatia* agglutinin (HPA, 1:50), peanut agglutinin (PNA, 1:10), and wheat germ agglutinin (WGA, 1:2000). For control the sections were preincubated in a solution with inhibitory sugars (Gal, GalNAc, GlcNAc), which resulted in marked reduction of the staining intensity.

4.4 Results

4.4.1 General findings (light microscopy)

Although mammary tissue was found in both genders the mammary glands are sexually dimorphic. Both females and males possess tubular scent glands.

4.4.1.1 Females

In the female the teat contains a relatively wide teat canal, which originates from a cisternal lactiferous sinus ("ampulla") with folded walls. This sinus is lined by an isoprismatic epithelium and gives rise to branching ducts (milkducts) which lead into lobules, which contain clusters of secretory acini. Sinus, ducts and acini are located under the cross striated sheet of skin muscle and are surrounded by connective and adipose tissue, containing blood and lymphatic vessels and nerves. This arrangement of the different components of the mammary gland generally corresponds to that of other mammals.

4.4.1.2 Males

In the male also a large sinus is to be found directly under the basis of the teat. The sinus is lined with cubic or prismatic epithelial cells. These glandular epithelial cells with apical protrusions apparently are active glandular cells and the lumen is filled with secretory material and cellular fragments. Glandular secretory units occur mainly in the connective tissue of the teat and differ morphologically from those of the female.

4.4.2 Specific histological and histochemical findings

4.4.2.1. Female mammary gland

The structures of the mammary gland of the female *Petrodromus* (ID number 302/07) are found between cross striated dermal musculature and cross striated muscles of the thoracic and ventral body walls (Fig. 4.1). The overall morphology is that of a non-lactating gland.

The teat canal is distally lined by keratinized and proximally by non-keratinized stratified squamous epithelium. The non-keratinized squamous epithelium continues into a two layered isoprismatic epithelium which lines a labyrinth of cisternal glandular spaces under the teat representing the sinus lactiferous. These cisternal spaces closely resemble the lactiferous sinus of humans. The epithelium of the cisterns partially forms high folds so that the lumen is narrowed down and can become slit-like. The apical cell layer is formed by basophilic isoprismatic to flattened cells. The cells of the basal layer are pale, myoepithelial cells.

From these cisternal spaces long small ducts arise, the epithelium of which is relatively flat but also consists of two cell layers: a flat surface cell layer and a deep layer of myoepithelial cells. These ducts lead to glandular lobules, which consist of tightly packed secretory units. The secretory units are more or less oval acini (Fig. 4.2).

They are composed of prismatic glandular epithelial cells and a basal layer of myoepithelial cells. The acinar lumen is narrow but well recognizable.

Actin

The positive demonstration of actin by immunohistochemistry shows that the basal cells of the teat canal and milk sinus are myoepithelial cells (Figs. 4.3 and 4.4). The milk sinus is collapsed (anastomosed) which is typical for non-lactating mammae.

PAS reaction (neutral glycoproteins)

The apex of the glandular epithelial cells of the acini, milk ducts and cisterns of the mammary gland stained in weak to medium intensity.

Alcian blue (pH 2.5)

This stain demonstrates polyanions, e.g. mucins with electrical charges. The material within the lumen of cisterns, ducts and secretory units did not stain with Alcian blue. However, the Alcian blue stained a surprisingly large number of mast cells in the connective tissue of the skin and the glandular lobules.

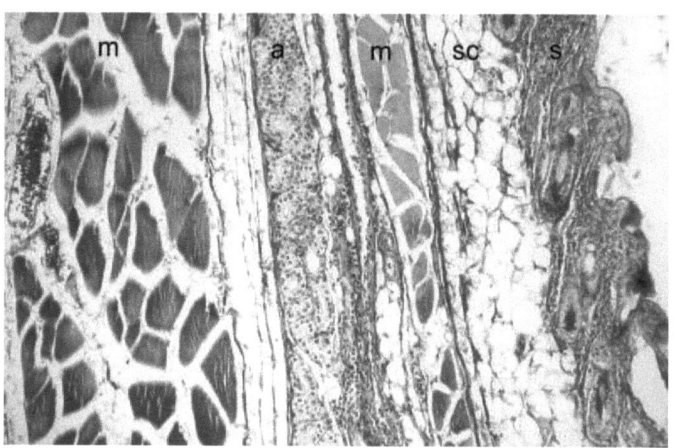

Fig. 4.1 *Petrodromus tetradactylus*, female, overview, dermal muscle (m), acini of mammary gland (a), subcutis (sc) and skin (s), Azan stain. Magnification x 120.

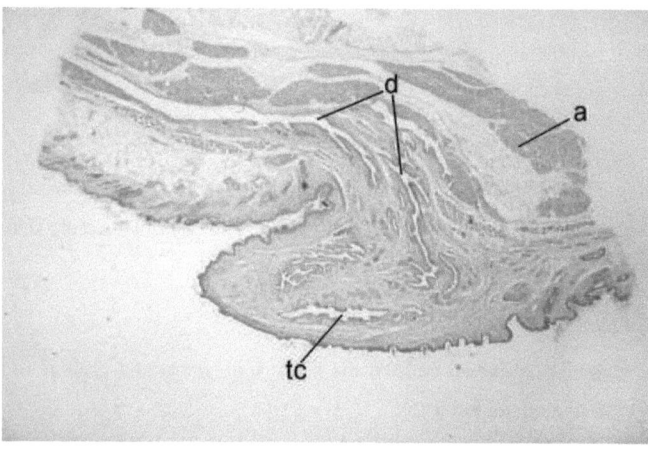

Fig. 4.2 *Petrodromus tetradactylus*, female, teat, acini (a), teat canal (tc) and ducts (d). Magnification HE x 20

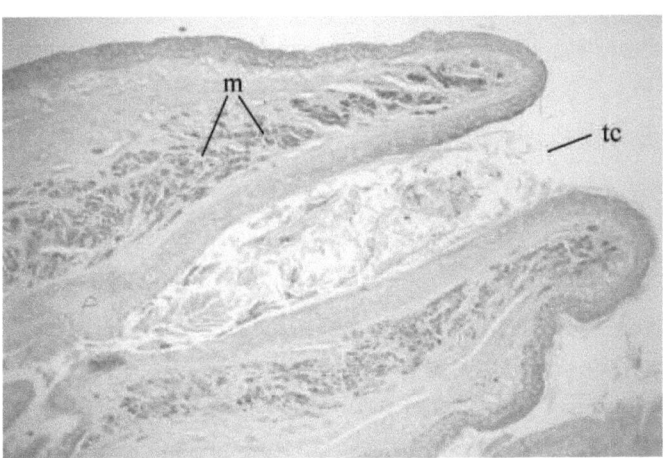

Fig. 4.3 *Petrodromus tetradactylus*, female. Demonstration of the presence of actin (brown stain) in the sphincter muscle (m) around the teat canal (tc). Magnification x 120.

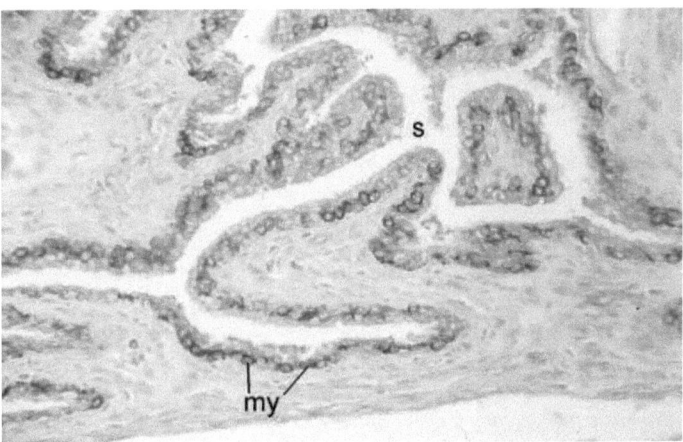

Fig. 4.4 *P. tetradactylus*, female. Demonstration of the presence of actin (brown stain) in the myoepithelial cells (my) of the anastomosed milk sinus (s). Magnification x 250.

Lectins

Lectin binding studies (WGA, HPA, and PNA) specify the results of the PAS reaction.

a) WGA (Wheat Germ Agglutinin)

A positive reaction of medium intensity was detected in the apex of the epithelial cells of the acini, the milk ducts and the cisterns. The stratified epithelium of the teat canal gave also a positive reaction of medium intensity, as did the entire epidermis. Also the myoepithelial cells were positive for WGA.

b) HPA (*Helix pomatia* Agglutinin)

Acini, milk ducts and cisterns of the mammary gland were – in contrast to the scent glands - negative for this lectin.

c) PNA (Peanut Agglutinin)

The epithelia of the acini, milk ducts and cisterns gave a positive reaction of medium intensity at the cellular apex. The reaction in the ductal and cisternal epithelia was slightly stronger than that of the acini. In addition, this lectin outlined clearly the capillaries surrounding the glandular tissue (Fig. 4.5).

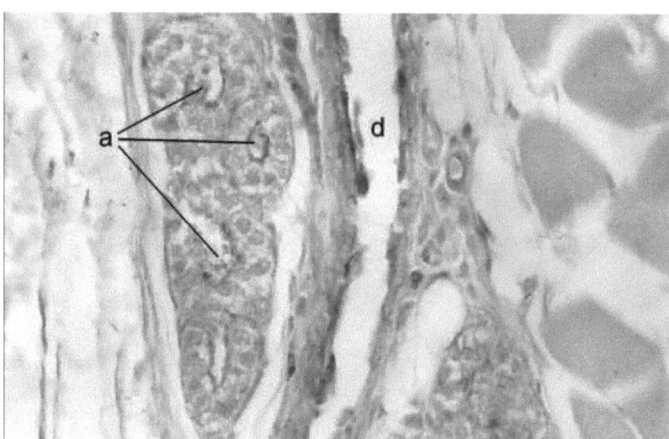

Fig. 4.5 *Petrodromus tetradactylus*, female. Binding of the lectin PNA (brown stain) at the apex of the acinar cells, thus outlining the lumen of the acini (a), small duct with positive reaction of the teat surface epithelium (d). Magnification x 450.

4.4.2.2 Male mammary gland

The teats of the male (ID numbers 216/07 and 217/07) also have a central excretory duct which distally consists of keratinized and proximally non-keratinized stratified squamous epithelium. The latter continues in the depth of the teat into a two layered isoprismatic epithelium which lines the wide cisternal space of a sinus lactiferous. The lumen of this cistern is filled with eosinophilic secretory products and necrotic cells which were probably sloughed off. The epithelial wall forms low and partially high folds. The surface epithelium is probably a secretory epithelium and the basal is a myoepithelium (positive immunostain for actin). No typical milk ducts have been observed radiating from the sinus. Acini occur in large numbers in the connective tissue of the teat (Figs. 4.6 and 4.7). These acini differ in morphology from those of the female: they are small and their lumen varies between narrow and comparatively wide. Single glandular cells contain large lipid inclusions. The glandular epithelium is mostly prismatic but can also be flattened.

Lectins

While PAS and Alcian Blue are negative in the sinus and acini, the lectin binding studies with WGA, HPA and PNA showed in part positive results for carbohydrates.

a) WGA Wheat Germ Agglutinin

The glandular cells of the acini in the teat stained in medium intensity – usually the entire cells were stained. The epithelia of the sinus were negative. However, cells and cell debris in the lumen of the sinus were positive for this lectin. The glandular cells of the scent glands showed a positive reaction at the apical membrane in medium intensity (Fig. 4.8).

b) HPA Helix pomatias agglutinin

The abundant acini in the teat were negative. The epithelial cells of the sinus gave a positive reaction at the apical membrane. In addition cells and cellular debris in the lumen of the sinus stained in medium intensity. The scent glands gave only a very weak positive reaction.

c) PNA Peanut Agglutinin

Acini and sinus of the mammary gland were negative for this lectin.

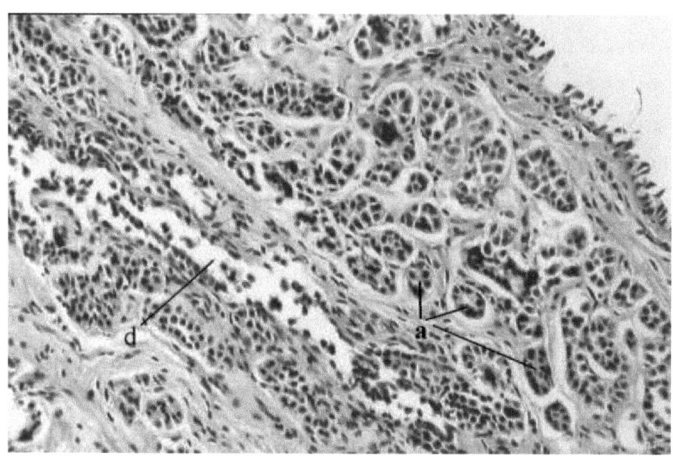

Fig. 4.6 *Petrodromus tetradactylus*, male, teat, acini (a) and duct (d) in the connective tissue of the teat. HE magnification x 250.

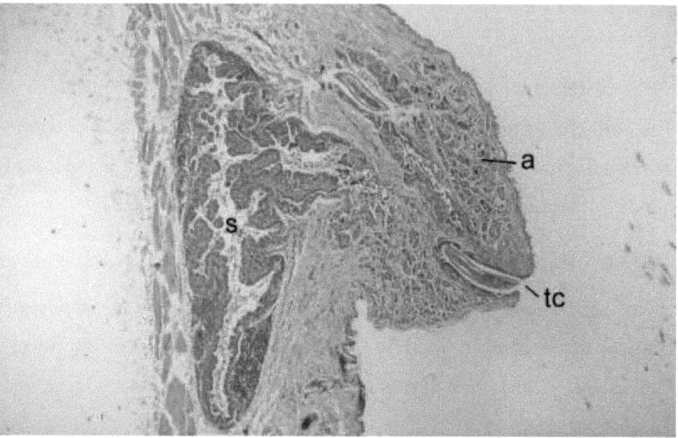

Fig. 4.7 *P. tetradactylus*, male, mammary gland, sinus (s), acini (a) in the connective tissue of the teat and opening of the teat canal (tc) where probably a small nematod is visible, HE magnification x 40.

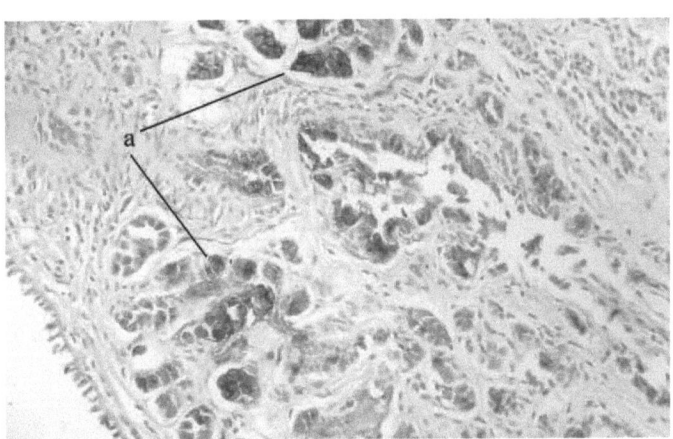

Fig. 4.8 *Petrodromus tetradactylus*, male, mammary acini (a) with positive WGA binding (brown stain). Magnification x 250.

4.4.2.3 Scent glands

The scent glands of both sexes are coiled tubular glands with a wide lumen (Figs. 4.9 and 4.10). The glandular epithelium is composed of prismatic epithelial cells, often with an apical protrusion, and myoepithelial cells. The glandular cells are active cells with morphological evidence for apocrine secretion in females (formation of an apical protrusion). These cells are characterized by a strong binding of the lectins WGA and HPA which stain the entire cytoplasm. PNA, however, was negative in contrast to the mammary gland.

In males WGA and HPA stained only the cellular apex in weak to medium intensity; PNA was also negative in the male.

Using the PAS method the entire cytoplasm of the apical protrusions in the scent glands gave a faintly positive stain, with individual cytoplasmatic granules staining strongly.

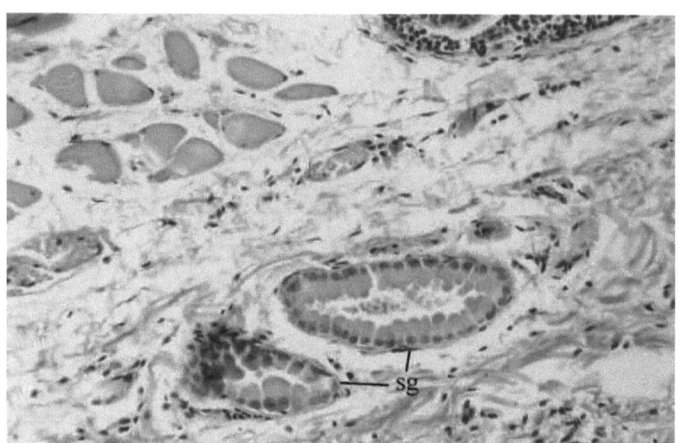

Fig. 4.9 *P. tetradactylus*, male, scent glands (sg), HE x 250.

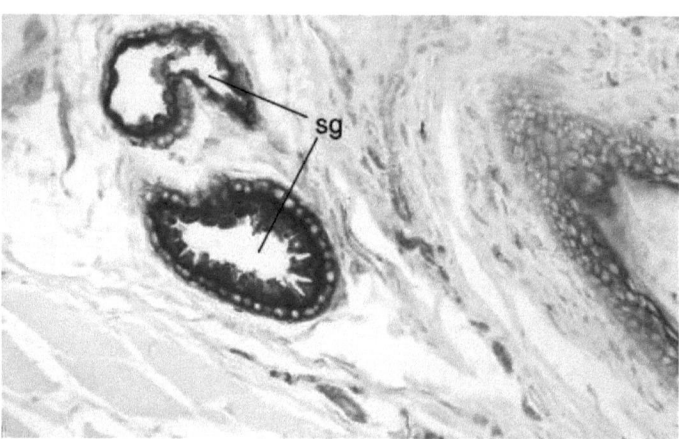

Fig. 4.10 *P. tetradactylus*, female, scent glands (sg) and positive reaction for HPA. Magnification x 250.

4.5 Discussion

4.5.1 Histo-morphology of mammary glands in *Petrodromus*

The present study has shown that not only female but also male *Petrodromus* are in possession of mammary glandular tissue. Histological techniques such as the hematoxylin-eosin stain and Heidenhain's azan stain demonstrate beyond doubt the presence of a differentiated gland with acini and ducts. In the female the histo-morphology closely resembles that of a non-lactating gland of other female mammals (Welsch et al. 2007). The well developed myoepithelial cells, as revealed by anti-actin immunohisto-chemistry underlines the close similarity. The demonstration of carbohydrates at the apex of the mammary epithelial cells (positive PAS-reaction, binding of WGA and PNA) indicates the presence of a differentiated glycocalyx which also occurs in non-lactating glands of humans (Welsch et al. 1998). The presence of PNA is an indication for secretory phenomena in the female in this study but there was no positive reaction for the male.

The situation of the male mammary gland of *Petrodromus* offers surprising aspects. Undoubtedly glandular acini are present, however, they are in a different location and of different histology if compared with those of the female specimen. The acini occur abundantly in the connective tissue of the teat and are smaller than those of the female. The acini show some indications of secretory activity. Several glandular cells have lipid droplets in their cytoplasma. Secretory material was present in both the male and female gland which was also found in the non-lactating mammary gland of a female human (Welsch et al. 2007).

As is the case in females, myoepithelial cells occur in acini of males. A common morphological feature of the female and male mammary gland is a wide cisternal space at the basis of the teats. This space corresponds to the milk-sinus of primates (Bargmann 1977). While in the female the glandular epithelium of this space had the appearance of a resting mammary gland, in the male its glandular cells were of variable height and several cells fragments were to be found in the lumen. This could indicate either an apocrine type of secretion or a process of a partial involution.

4.5.2 The presence of mammae in male sengis

Both in males of the bat *D. spadiceus* (Francis et al. 1994) and in *P. tetradactylus* the alveoli of the mammary tissue were more distended than in the females and in both species secretory material was found. In contrast to the fruit bats males of different sengi species have never been observed contributing to the care of young (Sauer 1973, Rathbun 1979b, Baker et al. 2005) and therefore, lactating males are not to be expected.

However, it is surprising that males of only two out of the eight sengi in this study, namely *P. tetradactylus* and *Elephantulus rozeti*, have evidence of mammae (Olbricht and Stanley subm.). Assuming that only these two species out of a total of all 16 sengi species are exceptional in having teats in males, it can only be speculated whether the existence of mammae is a relict due to phylogenetic inertia or whether it points to the high evolutionary potential of the Macroscelidea.

4.5.3 Other glands

Ancestral features of mammalian lactation are the basis for all other future modifications. In male *P. tetradactylus*, although not a basal macroscelid, not only mammary glands but also corresponding scent glands are found. Scent glands as well as mammary glands are apocrine glands and have their origin in sweat glands. Apocrine glands are considered to be the most primitive of the two types of scent glands (Quay 1977). Long (1969, 1972) discussed earlier hypotheses on the origin of lactation such as suckling of sweat in ancestral mammals that eventually led to a hypertrophied gland and subsequently to lactiferous structures. He concluded that placental viviparity rendered the pouch unnecessary but mammary glands persisted and developed teats. This theory would support the hypothesis that due to phylogenetic inertia teats in male *Petrodromus* persisted.

Other apocrine glands were found in both genders in a study of *Macroscelides proboscideus* and five species of *Elephantulus* (Faurie 1996) who stressed the importance of scent communication for sengis. A large number of actively secreting gland structures of the polyptych (holocrine, exocrine) type with neighbouring monoptych (endocrine) glands were found in the head and anal region of male *M. proboscideus* (Montag and Zeller 2001). The apocrine scent glands at the teat in *Petrodromus* are certainly important for the production of breast odour which helps newborns to find their mother's teat. This is well known for other mammals (Brown and Macdonald 1985). It remains unclear why scent glands are present at the male's teat. However, the importance of apocrine glands for sengis in general and the omnipotential of this type of glands may have resulted in the existence of male mammary glands with surrounding scent glands at least in two species.

Further, Daly (1979) stressed the fact that the refractoriness of the male mammary gland is relative rather than absolute. Although he states that physiological barriers to the evolution of male lactation do not seem individually insurmountable he concludes that "there is no indication that male lactation might ever occur spontaneously in natural mammalian populations." He further speculates that there might be a selective advantage in monogamous mammals if it ever got off the ground. And *Petrodromus* as well as the other sengi species are monogamous.

But these two theories are only speculative and the question of the purpose of mammmae in male *Petrodromus*, whether this is a relict of an ancestral feature or if it is a modification, cannot be answered here. The advanced features of lactation in a particular group are germane only for that group, and not of general applicability (Hayssen 1993). To ascertain the general applicability of a particular molecular or physiological process, its ancestry must be determined. However, *Petrodromus* is considered the most primitive species among the Macroscelidea (e.g. Evans 1942).

4.6 Conclusions

It is highly desirable to examine the teats of male *E. rozeti* where genetic and morphological research revealed a close relationship to *Petrodromus* (Douady et al. 2003). Their research reveals some evidence of possibly new arrangements of the members of the genus *Elephantulus* which was already proposed in 1912 by Heller who used the presence or absence of glands as taxonomic tools. Additionally, more research is needed to know more about the structures of monogamy in sengis which may allow to elucidate the hypothesis of the selective advantage of having males with functioning mammary glands.

CHAPTER 5

5. BODY METRICS OF THE SHORT-EARED SENGI (*MACROSCELIDES PROBOSCIDEUS*)

5.1 Abstract

Body measures (mass, length of head-body, ear, snout, whiskers, tail, hind foot) in altogether 36 males and 33 females of captive short-eared sengis (*Macroscelides proboscideus*) were measured post mortem. Body measures were then fitted to the 3-parameter Gompertz model. There was a considerable variation of the studied growth parameters in terms of growth constant (K) and inflection age (I). The whiskers and the snout had the fastest growth, the ears grew relatively slow. The asymptotic values of the growth model in terms of adult length of tail and ear as well as body mass were approached later then suggested by the sigmoidal curves but nevertheless, adult size of most body parts is achieved at about sexual maturity (ca. 45 days), except hind foot length which reached its maximum earlier. No significant sexual dimorphism in the estimated adult size could be determined.

5.2 Introduction

Growth pattern is an important life history trait (Stearns 1992). Assessment of approximate age relies on detection of age-related differences in characteristics such as body size. Age is one of the variables reflecting the structure and dynamics of a population. Knowledge on age and reproductive condition of individuals, together with data on sex ratio, is vital for evaluating the viability of a given population (Kunz et al. 1996). Consequently, established growth patterns may serve as a tool for surveys.

Although small mammals are less likely to produce mature neonates than large mammals there are exceptions on both sides (Derrickson 1992): Bears (e.g. *Ursus americanus*, Ewer 1973) are among the largest mammals with extremely altricial young; on the other hand, the short-eared sengi (*M. proboscideus*), as one of the smallest members of the Macroscelidea, is highly precocial.

5.2.1 Body metrics and other physical patterns

Body mass and size change rapidly during early stages of development and, therefore, are good indicators of age. The growth rate after birth is much better known than that in the prenatal period (Kunz et al. 1996). The dimensions of different body parts can be calibrated to age by monitoring the growth of known-age individuals.

In this study, standard measures were taken, as common in mammalogical practice: head-body or total length, tail length, hind foot length, ear length, and mass (e.g. DeBlase and Martin 1974, Melin et al. 2005, Kunz et al. 1996). Special attention is commonly paid to hind foot length as it is considered a diagnostic character (Krebs and Singleton 1993, Barko and Feldhamer 2002) at species and sub-species level and is important for the assessment of growth because it is less dependent of body size. However, Melin et al. (2005) argued that this standard metric is often taken routinely for its ease of aquisition. In short-eared sengis snout and whiskers are very evident body characteristics and therefore, were added to the measurement scheme.

The developmental stages of mammals have been generally examined based on the degree of epiphyseal fusion, tooth eruption and, sexual maturation (Shigehara 1980). The question arose whether maturity in sengis can be evaluated by the time when these three features are completed in their development. Recent studies revealed, that complete dentition in the Afrotheria is achieved much later than sexual maturity (Asher and Lehmann 2008). Asher and Lehmann (2008) concluded that sengis finish permanent dentition after they have reached adult body size. Further, the tendency for indeterminate growth was apparently present in both *Tenrec ecaudatus* and proboscideans. The attainment of adult body size in sengis prior to complete eruption of permanent teeth is not unique among mammals but it appears to depend on a combination of rapid growth and delayed eruption (Asher and Olbricht, subm.). This study was based on quantifications of skull metrics of *M. proboscideus* and *Erinaceus europaeus*.

5.2.2 Growth models

In addition to simple growth trajectories the growth of organism has frequently been modelled by sigmoidal functions. Many natural growth phenomena follow sigmoidal curves such as the logistic models, von Bertanlanffy´s and Gompertz´s (Zullinger et al. 1984).

The Gompertz model was chosen here, because it provides several growth parameters and is considered of particular use for comparative studies (Moscarella et al. 2001). In contrast to linear growth curves obtained by average growth rate (total weight gain divided by time in which it was achieved, cf. Millar 1977), growth models allow comparison among species with different body size. It it is the most widely used growth curve for rodents (e.g. Begall 1997, Begall and Burda 1998, Lammers et al. 2001, Jackson and van Aarde 2003).

5.2.3 Aim of the study

Data on absolute age of small afrotherians are lacking in scientific collections, so the aim of this study is to identify body size measures which could provide reliable correlates to age for field studies and scientific research. It may also contribute to monitor growth and development in captive sengis. Furthermore, standard growth parameters are established for comparative studies using other mammalian species.

5.3 Material and Methods

5.3.1 Material

The Zoological Garden Wuppertal, Germany, has bred short-eared sengis since 1988 and the breeding and longevity records in this location are exceptional (Olbricht 2007). Zoo data since the arrival of the first specimens are detailed and allow an insight into the natural history of this species over a long time period.

The 36 males and 33 females which were used in this study died of natural causes. The carcasses were frozen or preserved in 70% ethanol or 5% formalin. Measuring fresh or frozen and thawed carcasses was easier and provided more reproducible results than measuring wet conserved specimens. Some developmental stages are illustrated in Fig 5.1.

5.3.2 Methods

Body mass, head-body-length, tail length, hind foot length, ear length, nose length, and whisker length were measured post-mortem following the commonly used protocols (e.g., Harrison and Bates 1991). Measurements were taken to the nearest 0.1mm using a digital calliper and a string, body mass was measured before preserving the body to the nearest 0.01 g using an electronic scale. For ear length the external ear (pinna) was measured. The large somewhat folded ear had to be slightly stretched. Body length was measured using a string which was run dorsally along the body midline from the dorsal base of the proboscis-like nose over the head down to the base of the tail. Hind foot length was taken from the right foot. Measurements of the proboscis were taken ventrally from the base to the tip of the nose. Tail length was measured dorsally; the base was determined by lifting the tail up. The longest whisker which became visible under macroscopic view was extracted, placed straight on white paper and then measured.

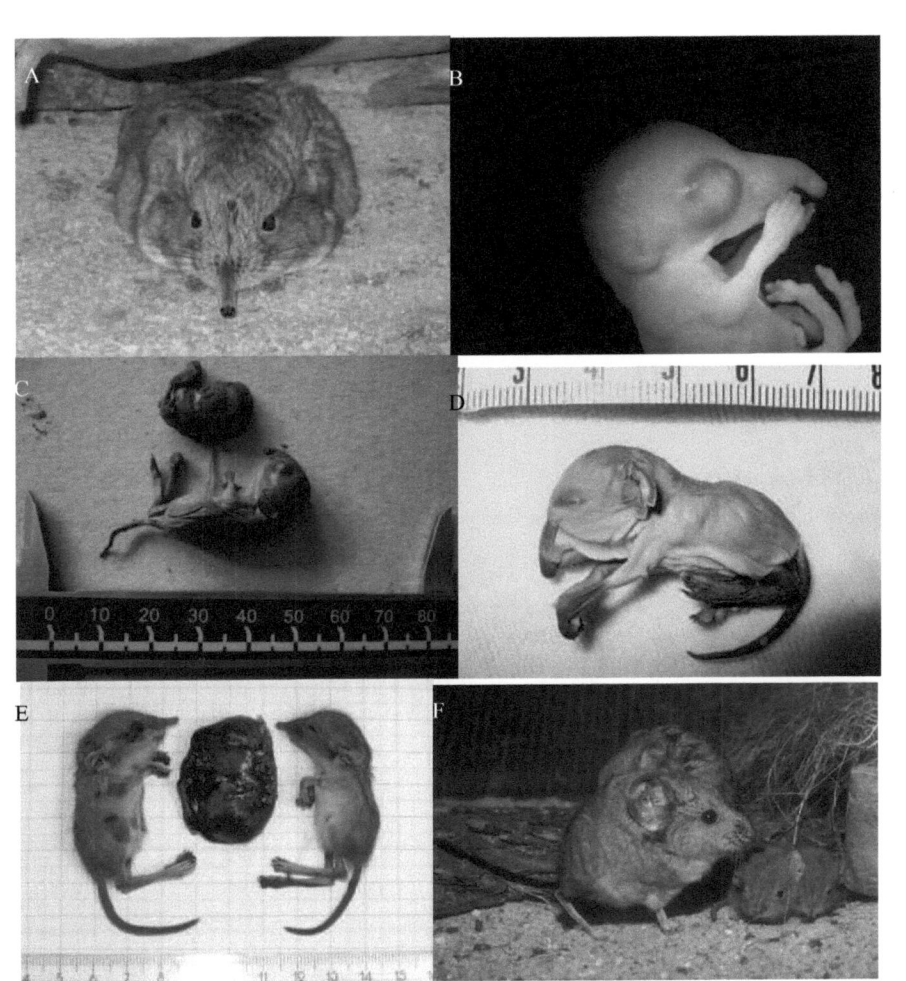

Fig. 5.1 Developmental stages of *M. proboscideus* (A-F, except B). (A) A pregnant female a few days before delivery attaining almost double of her body weight. (B) Very early fetus of unknown age of *Elephantulus spec.*, wild specimen Kimberley area, RSA, donated by the McGregor Museum, photo by S. Lange. (C) Early fetus of *M. proboscideus*. (D) Fetus, ca. 10 days before birth, (E) Newborn siblings with placenta, Zoo Leipzig, photo by C. Kern. (F) Female with young, a few days old, photo by A. Sliwa.

5.3.3 Statistical analysis

The Kolmogoroff-Smirnov-test was used to gauge whether the measured body parameters fit to normal distribution in adult animals. Since the data were normally distributed, parametric tests could be applied. I used a two-tailed t-test, including all data points (no means), to identify significant differences between adult females and males for each body parameter and body mass.

5.3.3.1 The Gompertz growth model

Data were fitted to a 3-parameter Gompertz model (cf., Begall 1997), using the following formula: $M(t) = A \cdot e^{-e^{-K(t-I)}}$ where m (t) = measured parameter at time t, A = asymptotic size, K = growth constant (day $^{-1}$), and I = inflection point (days). At the inflection point (I) the growth is the fastest. The asymptotic value (A) is an estimation of the parameter at adulthood (i.e. adult weight or adult size of the respective measured parameter).

Parameters were estimated using a non-linear regression procedure of SPSS.

5.4 Results

5.4.1 The Gompertz model

The growth parameters derived from the Gompertz equation (Table 5.1) demonstrate that there are distinctive variations within the growth of the various body parts and body mass.

Table 5.1 Mean growth parameter estimates for body mass, head-body, tail, hind foot, ear, snout and whisker, derived from Gompertz equations with A_{BM} = asymptotic value of the body mass (g), A = asymptotic value of measured parameters (except body mass) (cm); K = growth constant (days $^{-1}$) and I = inflection point (days). The 99% and 97% values are estimates for the age when 99% and 97% of the sample size reach the closest values to the asymptote A. At day A the asymptote is exceeded for the first time in the actual data set by at least one of the specimens. All other values are means calculated over all individuals.

Body parameter	A_{BM} (g) A (cm)	K (day $^{-1}$)	I (days)	Day A	99% of A (days)	97% of A (days)
Body mass	38.1	0.033	16.33	n. a.	155	122.13
Head-Body	10.05	0.053	- 5.731	49	81.06	60.14
Tail	9.73	0.072	0.551	113	64.4	49.04
Hind foot	3.18	0.078	- 11.03	38	47.9	33.73
Ear	2.25	0.031	- 9.96	101	138.4	102.66
Snout	1.02	0.097	- 1.9	18	47.2	35.8
Whisker	5.01	0.120	- 0.132	22	38.2	28.96

n. a. not applicable

The values for day A, when A was exceeded for the first time by one of the specimens, revealed that there are enormous differences in the time in which the maximum size of a particular body character is reached. While maximum snout length could be attained at day 18, adult tail length is reached for the first time at day 113.

These may be exceptions and to validate the data the age was calculated when most (97 % and 99 %) individuals reached adult size (Table 5.1). As expected, there were the same differences between the body measures as when calculating day A but here the earliest age to reach adult size was for whisker length with 29 days (97 %) or 38 days (99 %) and the latest for ear length with 103 days (97 %) and 138 days (99 %). However, while 99% of the specimens have reached adult size in whisker length by 38 days, ear length increases until day 138.

This corresponds to the growth constant (K) where whisker length has the fastest growth (K = 0.12) and ear length the slowest (K = 0.031).

The time of fastest growth (I) for five body characteristics out of seven, dates back to the pre-natal phase of the ontogeny (negative values). The earliest age at which maximum growth rate (I) was achieved, was for hind foot length with 11 days before birth and the latest age was for body mass with 16 days after birth. In Table 5.2 a ranking for I and K of all body parameters (taken from Table 5.1) is presented which demonstrates that the body measure with the earliest inflection point is not necessarily the one with the fastest growth. While hind foot, ear and head-body have the earliest dates for fastest growth, fastest growth in general occurs in the development of whiskers, snout and hind foot.

Table 5.2 The Gompertz parameters growth constant K (day $^{-1}$) and inflection point I (days) for each body parameter of *M. proboscideus* and their ranking with "1" for fastest growth or earliest inflection date and "7" for slowest growth and latest inflection date.

Body parameter	Growth constant (K)	Ranking for K	Inflection point (I)	Ranking for I
Ear	0.031	7	- 9.96	2
Whisker	0.120	1	- 0.132	5
Hind foot	0.078	3	- 11.03	1
Tail	0.072	4	0.551	6
Body mass	0.033	6	16.33	7
Snout	0.097	2	- 1.9	4
Head-Body	0.053	5	- 5.731	3

5.4.2 Sexual dimorphism

Seven body measures were recorded for male and female *M. proboscideus* (Table in appendix D), their mean growth trajectories are presented in Figs. 5.2-5.8.

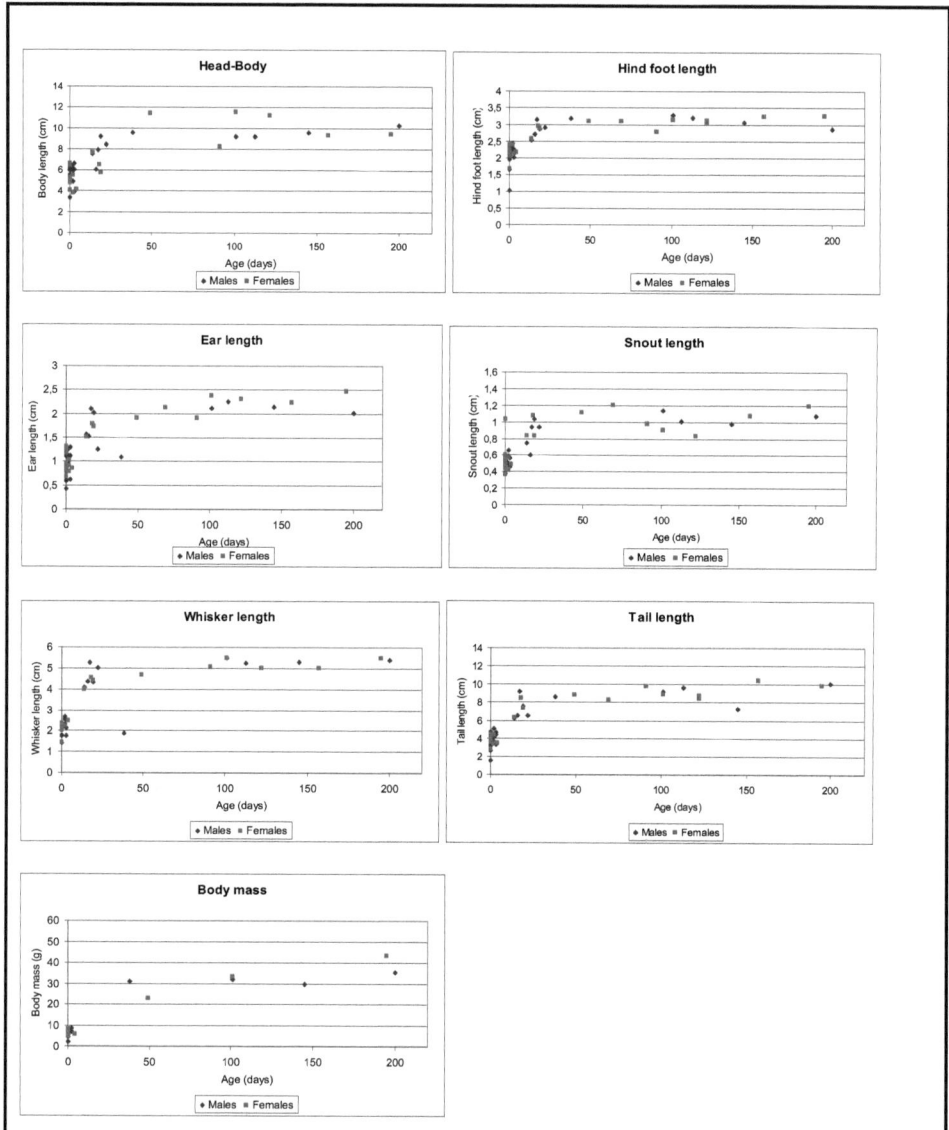

Fig. 5.2 – 5.8 Length of head – body, hind foot, ear, snout, whisker, tail and, body mass of male and female *M. proboscideus* in relation to age. Each data plot relates to a different specimen.

Notably, and in contrast to the growth model, adult size of various body measures is approached by some individuals at about 25 days.

A t-test revealed that there were no significant differences in final size between males and females in either body mass, length of tail, whisker, snout, ear, hind foot, and head-body length. I did not observe any sex or age dependent differences in fur colour and no sex-specific ornaments or integumentary pigmentation.

The asymptotic value (Table 5.1) was used to determine adult size and to indicate at what age maximum size is reached approximately. Specimens which reached or exceeded this size were considered adult for further calculations. These specimens were used for further analysis of sex-specific differences, and mean values ± standard deviations (SD) were calculated. A two-samples t-test was used to compare mean values for males and females (Table 5.3). Body mass is not considered in this table because the data set was too small. I will present the ±results for body mass which are computed on a larger data basis in chapter 6.

Table 5.3 Body measures (mean adult size ± standard deviation) of adult male and female *M. proboscideus* (results of the t-tests are indicated by p-values).

Body measure	Males Mean±SD	Females Mean±SD	significance (p, all data)
Hind foot	3.16±0.16	3.19±0.09	0.48
Head-body	10.12±1.22	9.94±1.13	0.72
Snout	1.01±0.09	1.01±0.1	0.94
Tail	10.06±1.58	9.64±0.75	0.49
Whisker	5.18±0.37	5.03±0.59	0.44
Ear	2.21±0.16	2.3±0.18	0.26

5.4.3 Correlation of body length with hind foot length and body mass

In contrast to previous growth curves with sigmoidal trajectories the correlation between length of the head-body and hind foot as well as body length to body mass until adulthood can be best described by a positive linear curve (Figs. 5.9 and 5.10).

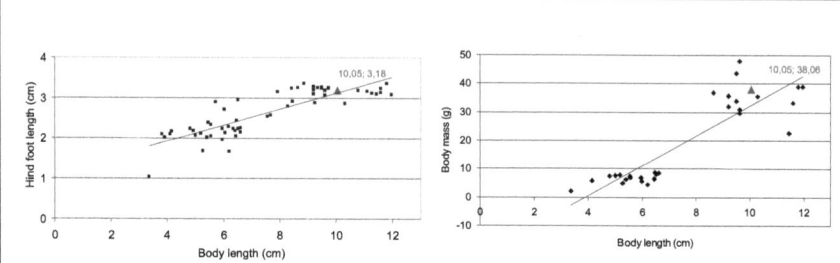

Fig. 5.9 and 5.10 Body length (head-body) versus hind foot length, and body length versus body mass of *M. proboscideus*. Linear regression lines were plotted into each graph. In red the asymptotes for these measures are indicated.

5.5 Discussion

5.5.1 Measuring scheme and statistical methods

The Gompertz model should be applied, when measurements from across the entire postnatal growth period can be incorporated (cf., Begall and Burda 1998). In this study, body measurements were not taken on a nearly daily basis but only once, on the date of death of an individual. All stillbirths were recorded as "day 1" but it is not clear whether this was a premature birth or whether possibly the newborn was already one or two days old and not found at birth. Birth weight may differ in sengis as is true for other mammals too (Naaktgeboren and Slijper 1970).

Modelling of other parameters reflecting growth has been successful. Specimens of different ages were available, and on this basis growth curves could be processed for *Macroscelides*, yet these results can only provide a rough estimate of growth for this species.

Furthermore, it must be taken into account that sickness, stress and suboptimal development as primary death causes may lead to a bias in body dimensions, especially in body mass. Consequently, this body measure within this data set must be judged critically since the animals were not in optimal body condition when they died. Body mass was not considered in some of the analyses because this parameter fluctuates daily or even hourly (Freeman and Jackson 1990), depending on gut content, weather and activity just to name a few. For very young neonates ingested milk can constitute a large proportion of body mass. An analysis of body mass growth is presented in chapter 6.

Maternal nutrition or body condition during early development also influences growth rates (Kunz et al. 1996). Kunz et al. (1996) emphasize that using body size to estimate the age of small animals in the field may be intriguing resulting from the difficulty of taking accurate linear measurements of live animals without anaesthesia. In this study a similar problem was found when measuring preserved animals such as frozen specimens or formalin or alcohol-fixed animals. This material varied in the quality of the preserved body, and sometimes I found it rather difficult to apply standard measurements. Here, bony structures such as hind foot length may be less influenced and thus are more favourable.

When comparing the results computing a percentage of 97 % with the percentage of 99 % of the accounts approaching the asymptotic value (Table 5.1), the 97 % estimate reveal a better approach to the real data (day A). To sum up the results, theoretical approaches and estimates processed by the Gompertz model mirror reality relatively well and serve as a useful basis for further research. Failures such as those discussed above are compensated for in an overall analysis.

5.5.2 Sexual dimorphism

In contrast to many rodent species (Lammers et al. 2001, Jackson and van Aarde 2003) sex-specific differences in growth could not be detected in *Macroscelides*. In contrast to *E. brachyrhynchus, E. rupestris* and *E. myurus* (Rautenbach and Schlitter 1977) sexual dimorphism in adult *Macroseclides* was not found either which is in accordance with adult *E. intufi* (Matson and Blood 1997). Sexual size dimorphism is strongly related to male mate competition and thus associated with a polygynous breeding system (Moscarella et al. 2001, Jackson and van Aarde 2003). Most sengis are monogamous mammals, and marked male-female body size dimorphism or other sex-specific differences are not likely to be needed for their breeding strategy (FitzGibbon 1995). Jackson and van Aarde (2003) also suggested that body size dimorphism may help shape local communities but sengis live solitarily or in pairs making sex-specific differences unnecessary. In contrast, mole rats are monogamous, but exhibit sexual dimorphism (Begall and Burda, 1998). Here are, however, overlaps as seen also in other traits with respect to the altricial/precocial dichotomy.

5.5.3 Developmental stage at birth and mating system

Some authors found a relationship between precociality and mating systems (Jackson and van Aarde 2003, Derrickson 1992), others found that species with altricial young tend to be monogamous (Zeveloff and Boyce 1986). Mole rats are altricial (Burda 1989) and sengis precocial, both are monogamous. Monogamy in sengis (Brown 1964, Woodall and Skinner 1989, Rathbun 1979a and 1981, Ribble and Perrin 2005, Rathbun and Rathbun 2006a) is expressed with low maternal investment so that precociality may be an advantage.

5.5.4 Growth patterns and maturity

5.5.4.1 The Gompertz growth parameters

Fastest growth occurs in the prenatal phase (I) which is true for the majority of mammals. Yet, there might also be a slight bias in the data which results from taking values of different animals into account instead considering one and the same individual over the complete (postnatal) growth phase (Begall 1997) and the problem of measuring specimens of *Macroscelides* which were preserved in fluid.

It may also be adaptive that precociality is linked to fast growth of important body parts. An example is the marsupial *Dasyurus quoll* where the forearm is much more developed at birth then the hind foot because it must climb on its own into the mother's pouch (Naaktgeboren and Slijper 1970). In the short-eared sengi growth constant K indicated that adult lengths of snout and whiskers are reached much earlier than those of other body parts. Here olfactory and tactile senses are located, which are crucial for the survival of the precocial young. Olfactory communication plays a particular role in the life of sengis (Faurie 1996, Montag and Zeller 2001). The flexible proboscis-like snout or nose is always in motion and assumingly plays an important role to detect food. Long whiskers aid tactile recognition as well. Well developed vibrissae are also found in neonate altricial rodents, i.e. *Fukomys anselli* as a subterranean mole-rat (Burda 1989b). Thus, vibrissae may be of particular importance for both precocial and altricial mammals, whereas tail length is not as important at this stage. Tail length in *Elephantulus intufi* and *E. rupestris* of various age classes exhibited greater variation than cranial measurements (Matson and Blood 1997).

Considerable skull growth in sengis takes place prior to attaining complete permanent dentition (Asher and Lehmann 2008) which may explain the early inflection point for head-body length. Asher and Lehmann (2008) also hypothesized that growth takes place faster and earlier in some afrotherians, e.g. sengis, than other mammals which could be related to the relative long gestation period. However, the authors stress that this is not true for Afrotheria as a whole clade. The earliest inflection age (I) of all body parameters was computed for hind foot length at 11 days before birth already which is certainly linked to the high importance of mobility for *Macroscelides* immediately after birth required for precocial neonates.

5.5.4.2 Body length versus hind foot length and body mass

Growth and maturity is usually assessed by means of age related ratios. I used correlations of body length versus hind foot length and body length versus body mass as a different approach. Burda (1989) concluded that weaning in *F. anselli* (rodent) depended primarily on the attained body size and not on the actual age.

The linear scaling of the three parameters body length, hind foot length and body mass included individuals comprising the entire ontogenetic series as suggested by Melin et al. (2005). The correlation of hind foot length with age had revealed that adult hind foot length is achieved very early whereas body mass increases over a long period of time. Sengis share this feature with other mammals such as bats (*Tadarida brasiliensis*) where Hermanson and Wilkins (2008) found advanced timing in hind limb bones (carpal and tarsal bones). Growth rate of hind foot length was greater in forefoot length in fetuses of the highly precocial nutria (*Myocaster coypus*, Sone et al. 2008). The inflection point of the hind foot length marks the time of weaning in the Norway rat (*Rattus norvegicus*) (Melin et al. 2005).

The regression curve of the correlation body mass versus body length was steeper than the regression curve for hind foot length versus body length.

Bivariate correlations indicate that hind foot length is a good predictor of overall body size but Freeman and Jackson (1990) suggest that fieldworkers might be cautious in their use of single external metrics. Relating hind foot length to head-body length or body mass Melin et al. (2005) found hind foot length in *Rattus norvegicus* the worst proxy for body size because the hind foot does not grow constantly during the post-natal period. Generally none of the metrics applied in their study scales linearly to head-body length or body mass. They conclude that differential growth rates in precocial mammals may be expressed differently.

5.5.4.3 Factors influencing precociality

The results of this study revealed that the inflection point for some body parts occurs in the prenatal phase. I infer that rapid and early physical development, even already before birth, may facilitate precociality. The differences in the ranking between K and I for the various body measures (Table 5.3) confirm the findings of Case (1978) and Eisenberg (1981) that the selective factors regulating postnatal growth rates are different from the factors determining the degree of precociality at birth. Derrickson (1992) pointed out that the time of birth in relation of development has been modified often during mammalian evolution and that the shift in timing may affect changes in size and life history traits. Variation in neonatal maturity among birds is associated with variation in physiological anatomical and behavioural traits (Nice 1962). For example precocial birds grow at only 25% of the rate of altricial species (Case 1978b). The semi-precocial bush Karoo rat (*Otomys unisulcatus*) undergoes rapid postnatal development and contrary to expectations displays K-selected reproductive characteristics (Pillay 2001). Both patterns are comparable with sengis.

5.5.4.4 Lactation period

Postnatal development relates among others to the duration of the lactation time. Melin et al. (2005) argue that in precocial mammals weaning and the attainment of independent coordinated locomotion are decoupled, so differential growth rates, if they exist at all, may be expressed quite differently. The duration of lactation of wild *Rhyncocyon chrysopygus* was 14 days, in *E. rufescens* 22 days (Rathbun 1976, 1979) and of captive *R. petersi* 45 days (Lengel 2007). I recorded one individual *Macroscelides* in captivity (not included in the data set) which lived independently from its mother at an age of 5 days, so that weaning age is normally at a later time as survival away from the mother becomes possible. In the wild Sauer and Sauer (1972) observed a female *Macroscelides* offering half digested ants to her 5 day old young but suckling was observed until about 3 weeks of age.

Among rodents, the highly precocial guinea pigs (Künkele and Trillmich 1997) and many hystricomorph rodents are also known to be suckled for much longer periods than it is required for the survival of the young (Weir 1974). Other reasons than physiological constraints, e.g. mother-offspring bonding or immunology are here of importance.

However, weaning of sengis is much earlier than in less developed small rodents such as *Spalacopus cyanus* (ca. 68 days) and altricial small mammals such as *Fukomys anselli* (formerly known as *Cryptomys hottentotus*: 72-105 days, Burda 1989b; 84 days, Begall et al. 1999). In contrast, hyraxes are suckled for as long as about 6 months (Olds and Shoshani 1982) which is much longer than one may expect for precocial mammals. Notably, the body mass of a neonate sengi is about 25% of the mother's body weight whereas in yellow-spotted hyraxes it is only about 10% but in some bats it is about 20% (Millar 1977). Millar (1977) concluded in his study on 100 mammal species that the size of a mammal at weaning, or independence, reflects the extent to which it is prepared for coping with the adult niche.

Burda (1989b) speculates in his study on rodents that length of development is primarily determined by phylogenetic factors and body size and that there is no general effect of the habitat upon length of gestation. However, *Rhynchocyon* has a shorter gestation period than *Macroscelides* and produces the least developed young among all other sengi genera. *Rhynchocyon* is also considered the most ancient genus among the extant sengis (Douady et al. 2003). Here the habitat may have an effect and it can be speculated that *Rhynchocyon* can afford to rear altricial young in the safety of the dense forest, all the more it is the only species which builds a nest. If the hypothesis is followed that ability to give birth to precocial young is an evolutionary novelty in mammals (Hopson 1973, Case 1978a) then the habitat of *Macroscelides* which lives in dry regions may have had an effect on gestation length.

5.5.4.5 Skeletal growth and sexual maturity

Hind foot and head-body axis are skeletal parts of the body parameters and good indicators for adult size. Following the trajectories (Figs. 5.2 - 5.8) adult size in these parameters is reached earlier than the Gompertz growth parameters suggest. Modelling growth the asymptote was exceeded for the first time (day A) for hind foot at day 38 and or head-body length at day 49. In *E. edwardii* 95% of adult hind foot length is reached by 30 days of age (Dempster et al. 1992).

Nevertheless, the findings of this study support behavioural and physiological observations of sengis. The dispersal phase was noted for *Macroscelides* at about 18 to 21 days (Sauer and Sauer 1972). Sexual maturity is assumed to start at about 41 to 45 days after birth when young *Macroscelides* were rigorously chased off the parental enclosure (Rosenthal 1975). Puberty in *E. rufescens* (Rathbun et al. 1981) usually occurs at two to three months of age but sexual maturity might be achieved with 50 days post partum already. Histological examination of the reproductive tract revealed for *E. myurus jamesoni* sexual maturity at about four to five weeks of age (van der Horst 1946) and investigations on teat development at five to six weeks of age (Mc Kerrow 1954).

Some body parameters might postdate their growth by a wide margin. Complete eruption of permanent dentition takes place much later than adult size and sexual maturity in sengis (Asher and Lehman 2008) and this is true for some body parameters such as length of ear and tail.

5.6 Conclusions

The Gompertz growth model is a tool to determine growth parameters in *Macroscelides* despite a fragmentary measuring scheme but can only present a trend due to the lack of complete data sets over a long period of the ontogenetic development of an individual. More complete data sets for body metrics, in particular on body mass, are needed to test the partially ambiguous results of the study comparing growth parameters derived from a model with growth trajectories based directly on the measures. A more detailed analysis based on a larger data set of body mass development of *M. proboscideus* is presented in chapter 6.

CHAPTER 6

6. POST-NATAL BODY MASS DEVELOPMENT IN *MACROSCELIDES PROBOSCIDEUS*

6.1 Abstract

Ten captive male and nine female short-eared sengis (*Macroscelides proboscideus*) were weighed on a regular basis from their first days of life until adulthood. The Gompertz growth model was applied to generate the growth parameters K, I and A which were compared with data on the reproductive biology. Furthermore, the growth parameters for *Macroscelides* were compared with those of various species obtained from the literature.

Adulthood is reached when adult size matches with sexual maturity, at about 45 days after birth. There were no significant differences between males and females in growth parameters or adult body mass.

6.2 Introduction

Body mass and its development is an important growth measure, and it is widely used to compare body growth of various species (Pontier et al. 1989, Kunz and Robson 1995, Cheng and Lee 2002, Barrionuevo et al. 2004). Grand (2005) stressed the importance of non-invasive measures, such as body mass and length, to increase improve our understanding of exotic species. Data on body mass development based on the data set of chapter 5 did not completely clarify to what extent survival independently from the mother, weaning, reaching adult size of various body measures and sexual maturity are related to each other. Consequently, it is extremely difficult to determine adulthood in the short-eared sengi.

Given that the degree of development are closely associated to the mating system (Derrickson 1992) and mating systems are linked to sexual dimorphism (Jackson and van Aarde 2003) it was also of interest to investigate sex specific growth patterns.

6.2.1 Growth models

Mammalian body growth is not linear, and fitting inappropriate models to measured data leads to a bias of estimates of growth rates (Zullinger et al. 1984). In their study on body mass increase of 331 species of mammals, Zullinger et al. (1984) found that the Gompertz model always provided the best or next best fit compared with the logistic and von Bertalanffy models. When applying growth models, Zullinger et al. (1984) generally pointed out the importance of large data sets to fit sigmoidal growth curves because it increases the statistical confidence in estimates of growth rates. Growth data cannot be fitted reasonably when the data are restricted to early growth, often prior to the inflection point or showing continued slow growth in adulthood with no clear asymptote. Consequently, information is lost when the data which are used to estimate growth rate are arbitrarily restricted.

6.2.2 Aim of the study

The results for body mass development obtained in chapter 5 were based on a small data set. The analysis of a larger data set used in this chapter refines the results of the previous chapter regarding body growth during the ontogeny of individual short-eared sengis. Additionally, sex specific growth patterns for body mass were examined.

6.3 Material and Methods

6.3.1 Material

Ten male and nine female short-eared sengis born at Wuppertal Zoo, Germany, were captured and weighed almost every day, some of them from the first day of their life. When twins of the same gender were born, one of them was marked by shearing off some fur on its back. All sengis were kept under nearly standardized husbandry conditions. Three females were hand-raised from their first or second day of life and kept under the same husbandry conditions in my home. All of them adapted well and did not show signs of delayed physical or social development.

6.3.2 Methods

The time of the daily sampling was scheduled between 10 and 12 a.m. before feeding time by the keeping staff of Wuppertal Zoo. Since sengis are able to jump high, it was necessary to use a plastic box with a lid to carry each individual from its enclosure to the electronic balance (Sartorius). This box was also used to weigh the animals. The animals were accustomed to this daily routine (Fig. 6.1), so that they did not move inside the closed box. Body mass was measured to the nearest 0.01g.

Fig. 6.1 Data collection at Wuppertal Zoo. Catching a sengi (left), weighing routine (center), facility for *M. proboscideus* (right)

6.3.2.1 Gompertz growth model and statistics

Body mass data were fitted to the Gompertz equation (Begall 1997; Zullinger et al. 1984; and see previous chapter for details). A t-test was applied to test for significant differences between mean values for the A and K parameter of males and females.

6.4 Results

6.4.1 Growth curves

Body mass increase of *M. proboscideus* shows a sigmoidal curve (Fig. 6.2). The trajectories for body mass increase of the various specimens are rather similar in shape. The numbers are ID numbers from Wuppertal Zoo, meaning F for female and M for male. There is no obvious variation among males and females or between mother-reared and hand-reared individuals. A plateau is reached at about 40 days after birth. There are two flatter curves: male M204004G died at day 19 and male M204004J was paired with a female at five weeks of age and possibly was developing differently then others due to social stress.

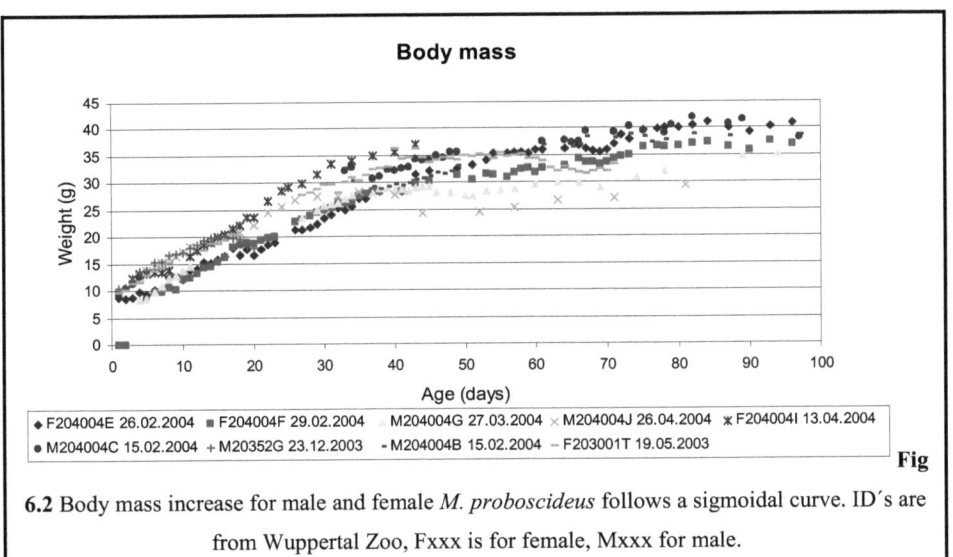

Fig 6.2 Body mass increase for male and female *M. proboscideus* follows a sigmoidal curve. ID's are from Wuppertal Zoo, Fxxx is for female, Mxxx for male.

6.4.2 Gompertz growth parameters

The growth parameters A (asymptotic value), K (growth constant) and I (inflection point) which were obtained for all specimens of this study, together with the period of the individual data collection, are listed in Table 6.1. Specimens included in Fig 6.2 are marked in bold and italics. Full data sets starting at the first three days of life until at least about three months after birth are marked in grey.

There was much inter-specific variation for all body mass parameters. The time of fastest growth (I) ranged from 18 days before birth (-18) up to 16 days after birth. The growth constant K varied between 0.013 days^{-1} and 0.19 days^{-1}, the estimated adult weight (A) between 20.08 g and 48.03 g. ID M20352G was also included in Fig. 6.2 and Table 6.1 (with K = 0.12 and A = 20.08 g) but to avoid bias it was not considered in further analyses because the data collection period was too short.

Table 6.1 Gompertz growth parameters A = asymptotic weight, K = growth constant; I = inflection point for female (Fxxxx) and male (Mxxxx) *Macroscelides proboscideus;* duration of data collection in days after birth; grey shading marks individuals with data set over a long period of time, bold and italics typeface for individuals included in Fig. 6.2; *: animal was not included in further analysis, hr: handraised, w: week.

ID	A (g)	K (day^{-1})	I (days)	Start/end data collect. (days)	Remarks
M203001Q	45.331	0.025	11.946	42/180	
M203001K	42.995	0.013	-18.39	50/195	
M203001O	42.154	0.024	-2.525	17/162	
*M20352G**	20.076	0.19	-1.688	0/19	died day 19*
M20352E	43.6	0.036	4.181	35/137	
M20352F	48.027	0.032	15.95	35/136	
M204004B	42.2	0.036	8.658	0/111	
M204004C	40.578	0.049	6.974	0/148	
M204004J	43.571	0.014	0.177	0/178	paired at 5 w
M204004G	33.236	0.055	4.259	0/90	
F203001V	37.181	0.026	11.435	10/505	lost tail
F203001T	36.41	0.062	7.291	2/147	
F203001P	46.1	0.027	10.567	23/394	
F204004H	43.005	0.016	7.573	6/158	
F204004I	47.836	0.046	16.37	6/48	
F205007D	43.589	0.032	12.753	3/72	sporadic data
F204004D	34.885	0.058	8.250	0/144	sick day 77, hr
F204004E	40.895	0.042	13.247	0/251	hr
F204004F	37.159	0.048	8.028	1/254	hr

6.4.2.1 Sex-specific growth

Mean asymptotic weight (A) for all males was 42.4 g (SD 4.04), for all females 40.7 g (SD 4.83) (Table 6.2). Mean growth constant (K) for males was 0.032 days^{-1} (SD 0.014) and for females 0.04 days^{-1}(SD 0.015). No significant differences were found between female and male *M. proboscideus* neither in adult body mass (p = 0.5 > 0.05) nor in growth constant K (p = 0.27).

The estimated adult weight (A) computed from the data set in chapter 5 (38.06 g) was much less than the mean asymptotes for both males and females, respectively, in this study. The mean growth constant of the specimens in chapter 5 was within the range of the data for male and female *Macroscelides* in this study (Table 6.1) but all K-values were smaller than that for for *E. rufescens* (Zullinger et al. 1984) which means that average growth was slower. in the *Macroscelides* of both studies (chapters 5 and 6). Nevertheless, seven out of the 18 samples of *M. proboscideus* (Table 6.1) had a similar or faster growth rate than the average of all individuals.

Table 6.2 Means of asymptotic weight (A) and growth constant (K).

Sample	Asymptote A (g)	Growth constant K (day^{-1})
Males	40.78	0.0315
Females	42.42	0.0396
Both sexes (chapter 5)	38.06	0.033
E. rufescens (Zullinger et al. 1984)	n.a.	0.0481

6.5 Discussion

In contrast to the growth curves (Fig 6.2) there was much inter-specific variation for all body mass parameters obtained by applying the Gompertz growth model. Case (1978b) computed the logs of weight increase (g/day) and did not employ the Gompertz model or other growth models as did Zullinger et al. (1984) but the K-values of both studies were similar.

Intraspecific variation may be the result of varying measuring intervals among the individual animals which confirms the importance of appropriate data sets. It can especially be troublesome if the choice of a time interval is restricted by patterns such as weaning. Weaning is about 21 days, the plateau for body mass growth is reached at about 40 days.

Weighing data were available for 7 individuals over a long period of time (grey shading in Table 6.1) but yet, there was much individual variation and no relation between "K" and "I" was recognizable. It is very likely that there is evidence for individual variations which is described also for several measurements of *E. intufi* and *E. rupestris* (Rautenbach and Schlitter 1977, Matson and Blood 1997).

6.5.1 Environmental and behavioural impact on weight development of study animals

Although the data collection was standardized as far as possible there was much intraspecific variation among individual development. Weight loss in the growing young can be caused by stress factors such as removing an offspring from its parental enclosure, antagonistic behaviour of parents towards their young or between siblings as well as introduction of a new breeding partner. An example is male M204004J that was paired with a female at 5 weeks of age already. The inflection point in his weight growth was earlier than in some others (0.18 days) but the growth constant was slower (0.014 day^{-1}). Its growth curve (Fig 6.29) was flatter but its estimated adult weight (A= 43.6 g) was higher than the mean.

Sometimes husbandry schemes such as keeper schedule changes, too large food items or suboptimal ambient temperatures affect the well-being of captive animals leading to lower growth rates. Additionally, the conditions experienced during early development can have profound long-term fitness effects and may lead to delayed maturity and reduced growth rates in cavies (Kraus et al. 2005).

Weight development and growth parameters varied among the hand-raised specimens but they generally did not differ significantly from the parameters and growth curves of mother-raised specimens.

6.5.2 Growth parameters

It is interesting to investigate the relationship between inflection age, growth constant and state of development at birth. The Brazilian free-tailed bat (*Tadarida brasiliensis*) has altricial young but exhibits more advanced development of hind limb bones than the Southeastern brown bat (*Myotis austroriparus*) which lives in the same environment (Hermanson and Wilkins 2008). Growth of *Tadarida* is characterized by a higher growth constant (0.112 day $^{-1}$) (Zullinger et al. 1984) than the *Macroscelides* of chapter 5 and 6 (from 0.0315 to 0.0396 day $^{-1}$), the inflection point is with 5 days earlier than for most *Macroscelides* specimens. In *Tadarida* no mother-offspring bonds exist but the young is reared by females of the colony which may reflect a strong constraint for fast development despite of altriciality.

The inflection point of *E. rufescens* at 11 days (Zullinger et al. 1984) ranges in the middle of the minimum and maximum obtained in this study for *Macroscelides* (range 0.18 days to 18.39 days). If one only considers "I" of animals with the complete data sets from about birth to adulthood, there is a range from 0.177 days to 13.247 days. On the base of the data set in the previous chapter the inflection point for body mass was at 16.33 days. For *E. rufescens* (Zullinger et al. 1984) the growth constant K for body mass (0.0481 day $^{-1}$) is in the middle range, if considering all individual *Macroscelides* (this study) but higher than the mean of the total data set of M*acroscelides*. I would suggest, that the large differences in the results for intraspecific "I" (this study) may reflect individual growth, caused by genetic or environmental factors. Variance in growth rate is known among precocial seals (Pinnipedia) (Case 1978b). At this point, the way of collecting of a complete data set of the same individual is not much likely to be of importance.

With regard to growth constant for body mass development of small mammals (Zullinger et al. 1984; Jackson and van Aarde 2003), *Macroscelides* is similar to members of the Rodentia (*Peromyscus, Meriones*), Lagomorpha (*Ochotona, Lepus*), Scandentia *(Tupaia, Microcebus)* and the Tenrecidae (*Echinops)*. Precocial mammals such as hares (*Lepus sp.*) have a range of "K" between 0.0191 and 0.0308 days^{-1}, *Macroscelides* is thus on the upper range. The range of "I" for *Lepus spec.* is between 39 and 58 days. (Zullinger et al. 1984) which is much later than *Macroscelides* (about 16 days maximum). In the highly precocial rodent *Cavia* has "K" = 0.0106 days^{-1} is related to "I" = 50 (*Cavia aperea*) or 61 (*Cavia porcellus*) days (Zullinger etl al. 1984).

Unfortunately, growth constant (K) and inflection date (I) cannot be related on an interspecific level on the base of Zullingers results and the results of this study because of different asymptotes. Neither could a relationship be established on an intraspecific level for *Macroscelides* resulting from the data of this study due to strong intraspecific variation. I agree that special attention has to be paid on the taxonomic level for which postnatal growth and period are measured (Pontier et al. 1989). Much interspecific variability is related to body size, yet substantial differences in growth rate often exist between species of similar adult size (Case 1978b). He concluded that in mammals, in general, there is no detectable difference in growth rate which may be attributed to the level of maturity at birth.

6.5.3 Adult body mass

6.5.4.1 Estimated adult weight (A)
Obesity in captive sengis was described for *Macroscelides* (Rosenthal 1975) and *E. rufescens* (Rathbun et al. 1981), but this was not an issue with the animals of this study. Rathbun et al. (1981) noticed that *E. rufescens* gained weight rapidly in captivity and maintained about 28.8 % more body mass compared to their weight right after capture in the field. The estimated adult weight and weight gain in general may be biased by this fact but Case (1978b) stated that superfluous feeding of mammals, e.g. under captive conditions, with species-specific growth should not enhance their growth rate over maximal rates observed in nature

Although the asymptotic weight (A) for *Macroscelides* based on the data set of chapter 5 (A = 38.06 g) is within the range for the individual specimens in this study (Table 6.1), the mean asymptote (A) calculated for larger data sets in this chapter predicts a larger body mass (42.4 g for males, 40.7 g for females).

The mean asymptotic weight was first approached in the actual data (Table D in appendix) at 79 days for males (M20352F with 42.74g) and, at 81 days for females (F204004E with 40.56g). Because this hand-raised female may have developed under different conditions, an additional account for females is listed here at 130 days (F203001P with 40.97g). However, the mean estimated body mass is reached far later than one may expect with regard to the growth curves in Fig. 6.2 where a tendency for the plateau becomes evident at about 45 days after birth which confirms the growth rates for *E. edwardii* (Dempster et al. 1992). Consequently, I infer that the

Gompertz model generally overestimates body mass, as already suggested previously for neonatal body mass (Zullinger et al. 1984).

6.5.4.2 Relating body mass and sexual maturity
Given that the lower range of adult body mass is achieved at about 45 days after birth (Fig. 6.2) this characteristic may be related to sexual maturity as suggested for other body measurements (chapter 5). *E. rufescens* reaches sexual maturity and adult weight at about 50 days after birth (Rathbun et al. 1981).

6.5.4.3 Sexual dimorphism
As I supposed in chapter 5, there is no sexual dimorphism, neither in body mass growth nor in adult body mass. Differences in weight between sexes are probably not adaptive in *Macroscelides*.

6.6 Conclusions

Body mass is a useful tool to model growth and to determine the onset of adulthood. Adult body mass relates to sexual maturity. Both methods, the post-mortem (Chapter 5) as well as the daily measurement scheme employed in this study, can be applied to growth rate analysis with the Gompertz equation. Modelling growth results in a number of parameters which are useful for inter- and intraspecific comparisons. Factors such as obesity, premature birth weights and preservation methods of dead specimens may lead to outliers but are compensated. Nevertheless, comparisons with other genera confirm Case's (1978) conclusions that there are no discernible differences in growth rate, which may be attributed to the level of maturity at birth.

IV REFERENCES

Alderton, D. 1996. Rodents of the world, Blandfort.

Altuna, C.A. and E.P. Lessa. 1985. Penial morphology in Urugayan species of *Ctenomys* (Rodentia: Octodontidae). J. Mammal. 66: 483-488.

Anderson, R.R. and K.N. Sinha. 1972. Number of mammary glands and litter size in the Golden hamster. J. Mammal. 53: 382-384.

Ansell, W.F.H. 1973. Mammals of the North-Eastern Montane Areas of Zambia. The Puku 7: 21.

Asher, R.J., M.J.N. Novacek and J.H. Geisler. (2003). Relationships of endemic African mammals and their fossil relatives based on morphological and molecular evidence. J. Mammal. Evol. 10: 131-194.

Asher, R.J. and T. Lehmann. 2008. Dental eruption in afrotherian mammals. BMC Biol. 6 (14): 1-11.

Baker, A.J., K. Lengel, K. McCafferty and H. Hellmuth. 2005. Black-and -rufous sengi (*Rhynchocyon petersi*) at the Philadelphia Zoo. Afrotherian Conserv. 3: 6-7.

Balter, M. 1997. Morphologists learn to live with molecular upstarts. Science 276: 1032-1034.

Bancroft, J.D. and A. Stevens. 1990. Theory and Practice of histological Techniques,:Churehill, Livingstone, Edinburgh.

Bargmann, W. 1977. Histologie und mikroskopische Anatomie des Menschen, Thieme Verlag, Stuttgart.

Barko, V.A. and G.A. Feldhamer. 2001. Cotton mice (*Peromyscus gossypinus*) in Southern Illinois: Evidence for hybridization with white-footed mice (*Peromyscus leucopus*). Am. Midl. Nat. 147:109-115, 2002.

Barrionuevo, F.J., F. Zurita, M. Burgos, and R. Jimenez. 2004. Developmental stages and growth rate of the mole *Talpa occidentalis* (Insectivora, Mammalia). J. Mammal. 85: 120-125.

Barry, R.E. and J. Shoshani. 2000. Heterohyrax brucei. Mammal. species 645: 1-7.

Begall, S. 1997. The application of the Gompertz model to describe body growth. Growth development & Aging 61: 61-67.

Begall, S. and H. Burda. 1998. Reproductive characteristics and growth in the eusocial Zambian common mole-rat (*Cryptoms sp.*, Bathyergidae). Z. Säugetierkunde 63: 297-306.

Begall, S., H. Burda and M.H. Gallardo. 1999. Reproduction, postnatal dvelopment, and growth of social coruros, *Spalacopus cyanus* (Rodentia: Octondidae), from Chile. J. Mammal 80: 210-217.

Bernard, R.T., G.H.I. Kerley, T. Dubell and A. Davison. 1996. Reproduction in the round-eared elephant shrew (*Macroscelides proboscideus*) in the southern Karoo, South Africa. J. Zool. Lond. 240: 233-243.

REFERENCES

Blüm, V. 1985. Vergleichende Reproduktionsbiologie der Wirbeltiere. Springer-Verlag,

Boitani, L.; F. Corsi, L. De Biase, I. D'Inzillo Carranza, M. Ravagli, G. Regiani, I. Sinibaldi and P. Trapanese. 1990. A databank for the Conservation and Management of the African Mammals. Europ. Commission, Directorate-General for Development, Division VIII/A/1 and Istituto Ecologia Applicata, Rome, Italy.

Brotherton P.N.M. and A. Rhodes. 1996. Monogamy without biparental care in a dwarf antelope. Proc. Biol. Sci. 263 (1366): 23-29.

Brotherton, P.N.M. and M.B. Manseri. 1997. Female dispersion and the evolution of monogamy in the dik-dik. Anim. Behav. 54: 1413-1424.

Brotherton, P.N.M., J.M. Pemberton, P.E. Komers and G. Malarky. 1997. Genetic and behavioural evidence of monogamy in a mammal, Kirk´s dik-dik (*Madoqua kirkii*). Proc.Zool.Soc.London 264: 675-681.

Brown, J.C. 1964. Observations on the elephant shrews (Macroscelididae) of Equatorial Africa. Proc.Zool.Soc.London 143: 103-119.

Brown, R.E. and D.W. Macdonald. 1985. Social odours in mammals, vol. 2. Clarendon Press, Oxford.

Burda, H. 1989a. Relationships among rodent taxa as indicated by reproductive biology. Zeitschrift für zool.Systematik und Evol.forsch. 27: 49-57.

Burda, H. 1989b. Reproductive biology (behaviour, and postnatal development) in subterranean mole-rats, *Cryptomys hottentotus* (Bathergidae). Z. Säugetierkunde 54: 360-376.

Butler, P.M. 1956. The skull of Ictops and the classification of the Insectivora. Proc .Zool. Soc. Lond. 126: 453-481.

Butler. P.M. 1995. Fossil Macroscelidea. Mammal Rev. 25: 3-14.

Carlsson, A. 1909. Die Macroscelididae und ihre Beziehungen zu den übrigen Insectivoren Zoologische Jahrbücher, Abt. Systematik, Ökol. und Geographie der Tiere 28: 349-400.

Carter, A.M. and A.C. Enders 2004. Comparative aspects of trophoblast development and placentation. Reprod. Biol. Endocrinol. 2: 46.

Carter, A, M., A. Enders, C.H. Künkel, D. Oduor-Okelo, D. and P. Vogel. 2004. Placentation in species of phylogenetic importance: the Afrotheria. Animal Reprod. Sci. 82-83: 35-48.

Case, T.J. 1978a. Endothermy and parental care in the terrestrial vertebrates. The Am. Natural. 112: 861-874.

Case, T.J. 1978b. On the ovulation and adaptive significance of postnatal growth rates in the terrestrial vertebrates. The Quarterly Rev. of Biol 53: 243-282.

Cheng, H.C. and L.L. Lee. 2002. Postnatal growth, age estimation, and sexual maturity in the Formosan leaf-nosed bat (*Hipposideros terasensis*). J. Mammal. 83: 785-793.

Clutton-Brock, T.H. 1988. Reproductive Success. In: (T. H. Clutton-Brock, ed) Reproductive Success. Chicago:Univ. of Chicago Press: 472-485.

Coetzee, C.G. 1966. The relative position of the penis in southern African dassies (Hyracoidea) as a character of taxonomic importance. Zoologica Africana 2: 223-224.

Coldiron, R.W. 1977. On the Jaw Muscuature and Relationships of *Petrodromus tetradactylus* (Mammalia, Macroscelidea). Am. Mus. Nov. 2613: 1-12.

Cooper, A.P. 1840. On the anatomy of the breast. Rare Medical Books. Thomas Jefferson Univ., Philadelphia.

Corbet, G.B. and J. Hanks. 1968. A revision of the elephant-shrews, family Macroscelididae. Bulletin of the British Museum of Nat. Hist. 16: 1-113.

Corbet, G.B. 1995. A cladistic look at classification within the subfamily Macroscelidinae based upon morphology. Mamm. Rev. 25: 19-30.

Daly, M. 1979. Why don't male mammals lactate? J. Theor. Biol. 78: 325-345.

de Graaf, G. 1981. The rodents of southern Africa. Butterworth and Co. Ltd. Durban, Pretoria.

de Jong, W.W., Zweers, A. and M. Goodmann. 1981. Relationship of aardvark to elephants, hyraxes and sea cows from alpha-crystallin sequences. Nature 292: 538-540.

Dempster, E.R., M.R. Perrin and R.J. Nuttal. 1992. Postnatal development of three sympatric small species of southern Africa. Z. Säugetierkunde 57: 103-111.

Derocher, A.E. 1990. Supernumerary mammae and teats in the polar bear (*Ursus maritimus*). J. Mammal. 71: 236-237.

Derrickson, E.M. 1992. Comparative reproductive strateges of altricial and precocial eutherian mammals. Funct. Ecol. 6:57-65.

Douady, C.J., F. Catzeflis, J. Raman, M.S. Springer and M.J. Stanhope. 2003. The Sahara as a vicariant agent, and the role of miocene climate events, in the diversification of the mammalian order Macroscelidea (elephant shrews). Proc. Natl. Acad. Sci. USA 100: 8325-8330.

Eisenberg, J.F. 1981. The mammalian radiations. The University of Chicago Press, Chicago, USA.

Eisenberg, J.F. and E. Gould. 1970. The Tenrecs. A study in mammalian behaviour and evolution. Smithsonian contributions to Zool. Smithsonian Inst. Press. City of Washington 27: 119-121.

Ellenberger, W. and H. Baum. 1974. Handbuch der vergleichenden Anatomie der Haustiere, Springer Verlag.

Erickson, A.W. 1960. Supernumerary mammae in the black bear. J. Mammal. 41: 409.

Evans, F.G. 1942. The osteology and relationships of the elephant shrews (Macroscelididae). Bull. Am. Mus. Nat. Hist. 80: 85-125.

Faurie, A.S. 1996. A comparative study of communication in six taxa of southern African elephant shrews (Macroscelidea). Pietermaritzburg, MS Thesis, University of Natal, RSA. pp. 22-47.

Fischer, M.S. 1986. Die Stellung der Schliefer (Hyracoidea) im phylogenetischen System der Eutheria. Courier Forsch.-Inst. Senckenberg, Frankfurt a. Main 84: 1-132.

FitzGibbon, C.D. 1995. Comparative ecology of two elephant-shrew species in a Kenyan coastal forest. Mamm. Rev. 25: 19-30.

Fitzinger, L. J. 1867. Über die natürliche Familie der Rohrrüßler (Macroscelides) und die derselben angehörigen Arten. Sitzungsberichte der Kaiserlichen Akademie der Wissenschaften 56: 2-28.

Fitzsimons, F. W. 1920. The natural history of South Africa. Brit. Mus. Nat. Hist. London.

Francis, C. M., E.L. Anthony, J.A. Brunton and T.H. Kunz. 1994. Lactation in male fruit bats. Nature 367: 691-692.

Freeman, S. and W. Jackson. 1990. Univariate metrics are not adequate to measure avian body size. The Auk 107: 69-74.

Frey, R. 1994a. Der Zusammenhang zwischen Lokomotionsweise, Begattungsstellung und Penislänge bei Säugetieren, I. Testiconda. Z. zool. Systematik und Evolutionsforsch. 32: 163-179.

Frey, R. 1994b. Der Zusammenhang zwischen Lokomotionsweise, Begattungsstellung und Penislänge bei Säugetieren, II. Testiphaena. Z. zool. Systematik und Evolutionsforsch. 32: 163-179.

Gibbons, A. 1991. Systematics goes molecular. Science 254: 872-874.

Gilbert, A.N. 1986. Mammary number and litter size in Rodentia: The "one-half rule". Proc. Natl. Acad. Sci. USA 83: 4828-4830.

Glover, T.D and Sale, J.B. 1968. The reproductive system of male rock hyrax (*Procavia* and *Heterohyrax*). J. Zool. 156: 351-362.

Grafen, A. 1988. On the Uses of Lifetime Reproductive Success. In: (T. H. Clutton-Brock, ed) Reproductive Success. Univ. of Chicago Press: 454-471.

Grand, T.I. 2005. Altricial and precocial mammals: a model of neural and muscular development. Zoo Biology 11: 3-15.

Haeckel, E. 1866. Generale Morphologie der Organsimen. Bd 1. Verlag Georg Reimer, Berlin.

Haltenorth T. and H. Diller. 1977. Säugetiere Afrikas und Madagaskars.. BLV Bestimmungsbuch. 1st(19): 314-320.

Hartman, C. G. 1927. A case of supernumerary teat in Macaca rhesus, with remarks upon the biology of polymastia and polythelia. J. Mammal. 8: 96-106.

Harvey, P.H. and R. M. Zammuto. 1985. Patterns of mortality and age at first reproduction in natural populations of mammals. Nature 315: 319-320.

Hayssen, V., A. van Tienhoven and A. van Tienhoven. 1993. Asdell's Patterns of mammalian Reproduction. USA:Cornell University Press, Ithaca, NY. Pp. 702-704.

Hedges, S.B. 2001. Afrotheria: Plate tectonics meets genomics. Proc. Natl. Acad. Sci .USA 98:1-2.

Heller, E. 1912. New cases of insectivores, bats, and lemurs from British East Africa. Smithsonian Miscellaneous Coll. 60: 1-10.

REFERENCES

Hennemann III., W.W. 1984. Intrinsic rates of natural increase of altricial and precocial eutherian mammals: the potential price of precociality. Oikos 43 (3): 363-368.

Hermanson, J.W. and K.T. Wilkins. 2008. Growth and development of two species of bats in a shared maternity roost. Cells, Tissues, Organs 187: 24-34.

Hoeck, H.N. 1977. "Teat order" in Hyrax (*Procavia johnstoni* and *Heterohyrax brucei*). Z. Säugetierkunde 42: 112-115.

Hopson, J.A. 1973. Endothermy, small size, and the origin of mammalian reproduction. The Am. Natural. 107: 446-452.

Hufnagl, E. 1972. Libyan mammals. Oleander Press, USA. pp 29-30.

Imperato-McGinley, J. Biniendo, Z. Gedney, J. and E.D. Jr. Vaughan. 1986. Teat differentiation in fetal male rats treated with an inhibitor of the enzyme 5 alpha-reductase: defintion of a selective role for dihydrotestosterone. Endocrinol. 118: 132-137.

Jackson, T.P. and R.J. van Aarde. 2003. Sex- and species-specific growth patterns in cryptic African rodents, *Mastomys natalensis* and *M. coucha*. J. Mammal. 84: 851-860.

Jennings, M.R. and G.B. Rathbun. 2001. *Petrodromus tetradactylus*. Mammalian Species 682: 1-6.

Jones, M.L. 1982. Longevity of Captive Mammals. Zool.Garten N.F. 52: 113-128.

Jones, W.T. 1985. Body size and life-history variables in heteromyids. J. Mammal. 66: 128-132.

Kaudern, W. 1910. Studien über die männlichen Geschlechtsorgane von Insectivoren und Lemuriden. Zool. Jahrbücher, Abt. für Anat. 31: 1-106.

Kerley, G.I.H. 1989. Diet of small mammals from the Karoo, South Africa. S. Afr. J. Wildl. Res. 19: 67-72.

Kerley, G.H.I. 1995. The round-eared elephant-shrew *Macroscelides proboscideus* (Macroscelidea) as an omnivore. Mamm. Rev. 25: 39-44.

Kim, S.W., R.A. Easter and W.L. Hurley. 2001. The regression of unsuckled mammary glands during lactation in sows: The influence of lactation stage, dietary nutrients, and litter size. J. Anim. Sci. 79: 2659-2668.

Kingdon, J. 1984. East African Mammals. The University of Chicago Press. 2nd ed. (II Part A): 37 pp.

Kingdon, J. 1990. Island Africa. Collins, London, UK.

Komers, P.E. 1996. Obligate monogamy without parental care in Kirk's dikdik. Anim. Behav. 51: 131-140.

Komers, P.E. and P.N.M. Brotherton. 1997. Female space use is the best predictor of monogamy in mammals. Proc. Zool. Soc. London 264: 1261-1270.

Koontz, F.W. and N.J. Roeper. 1983. *Elephantulus rufescens*. Mammalian species 204: 1-5.

Kraus, C., F. Trillmich and J. Künkele. 2005. Reproduction and growth in a precocial small mammal, *Cavia magna*. J. Mammal. 86: 763-772.

Krebs, C.J. and G.R. Singleton. 1993. Indexes of condition for small mammals. Australian Journal of Zoology 41: 317-323.

Kunz, T.H. and S.K. Robson. 1995. Postnatal growth and development in the Mexican Free-tailed bat (*Tadarida brasiliensis*): birth size, growth rates, and age estimation. J. Mammal. 76: 769-783.

Kunz, T.H., C. Wemmer and V. Hayssen. 1996. Sex, age, and reproductive condition of mammals. In: (D. Wilson, F.R. Cole, J.D. Nichols, R. Rudran, and M.S. Foster, eds) Measuring and monitoring biological diversity. Smithsonian Inst. Press, Washington and London. pp. 279-290.

Künkele, J. and F. Trillmich. 1997. Are precocial young cheaper? Lactation energetics in the guinea pig. Physiol. Zool. 70: 589-596.

Künkele. J. 2000. Effects of litter size on the energetics of reproduction in a highly precocial rodent, the Guinea pig. J. Mammal. 81: 691-700.

Lammers, A.R., H.A. Dziech and R.Z. German. 2001. Ontogeny of sexual dimorphism in *Chinchilla lanigera* (Rodentia: Chinchillidae). J. Mammal. 82: 179-189.

Lawes M.L. and M.R. Perrin. 1995. Risk-sensitive foraging behaviour of the round-eared elephant shrew (*Macroscelides proboscideus*). Behavioural Ecology and Sociobiol. 37: 31-37.

Leirs, H., H. Verhagen W. Verheyen and M. R. Perrin. 1995. The biology of Elephantulus brachyrhynchus in natural miombo woodland in Tanzania. Mamm. Rev. 25: 45-50.

Lengel, K. 2007. American Regional Studbook for the giant elephant shrew. Philadelphia Zoo.

Lidicker, W.Z. and P.V. Brylski. 1987. The conilurine rodent radiation of Australia, analyzed on the basis of phallic morphology. J. Mammal. 68: 617-641.

Long, C.A. 1969. The origin and evolution of mammary glands. Bioscience 19: 519-523.

Long, C.A. 1972. Two hypotheses on the origin of lactation. The Am. Natural. 106: 141-144.

Louwman, J.W.W. 1973. Breeding the tailless tenrec at Wassenaar Zoo. Int. Zoo Yearb. 13: 125-126.

Lovegrove, B.G., J. Raman and M. R.Perrin. 2001a. Heterothermy in elephant shrews, *Elephantulus spp*. (Macroscelidea): daily torpor or hibernation? J. Comp. Physiol. B 171: 1-10.

Lovegrove, B.G., Raman, J. and M. R. Perrin. 2001b. Daily torpor in elephant shrews (Macroscelidea: *Elephantulus spp.*). J. Comp. Physiol. B 171: 11-21.

Lowery, Jr., G.H. 1981. The mammals of Louisiana and its adjacent waters, Louisiana State Univ. Press.

Lumpkin, S., F. Koontz and J.G. Howard. 1982. The oestrous cycle of the rufous elephant shrew *Elephantulus rufescens*. J. Reprod. Fert. 66: 671-673.

Lumpkin, S. and F. W. Koontz. 1986. Social and sexual behavior of the Rufous elephant shrew (*Elephantulus rufescens*) in captivity. J. Mammal. 67: 112-119.

Mares, M.A. 1988. Reproduction, Growth, and Development in Argentine gerbil mice, *Eligmodontia typus*. J. Mamm. 69: 852-854.

Martin, R.D. and A. M. MacLarnon. 1985. Gestation period, neonatal size and maternal investment in placental mammals. Nature 313: 220-223.

Martin, R.E., Pine, R.H. and A.F. DeBlase. 2001. A manual of mammalogy, with keys to families of the world. McGraw Hill, Boston, Massachusetts.

Matson, J.O. and B.R. Blood. 1997. Morphological variability and species limits in elephant shrews (*Elephantulus intufi and E. rupestris*) from Namibia. Mammalia 62: 77-93.

McKenna, M.C. 1975. Toward a phylogenetic classification of the Mammalia. In: (W.P. Luckett and F.S. Szylay, eds) Phylogeny of the primates. Plenum Press, New York. pp. 21-46.

McKerrow, M.J. 1954. The lactation cycle of Elephantulus myurus jamesoni (Chubb). Phil. Trans. R. Soc. Lond. 238: 62-98.

Melin, A.D., P.J. Bergmann and A.P. Russell. 2005. Mammalian postnatal growth estimates: the influence of weaning on the choice of a comparative metric. J. Mammal. 86: 1042-1049.

Mess, A. and A.M. Carter. 2006. Evolutionary transformation of fetal membrane characters in Eutheria with special reference to Afrotheria. J. Exp. Zool. 306B: 140-163.

Millar, J.S. 1977. Adaptive features of mammalian reproduction. Evolution 31: 370-386.

Mohr, M. 1965. Altweltliche Stachelschweine. Die Neue Brehm-Bücherei.

Montag, M. and U. Zeller. 2001. Die Duftdrüsen der kurzohrigen Elefantenspitzmaus, *Macroscelides proboscideus*. in: 75. Jahrestagung der Dt. Ges. für Säugetierkunde, Berlin. Dt. Ges. für Säugetierkunde:26-27.

Moscarella, R.A., M. Benado and M. Aguilera. 2001. A comparative assessment of growth curves as estimators of male and female ontogeny in *Oryzomys albigularis*. J. Mamm. 82: 520-526.

Neal, B.R. 1982. Reproductive ecology of the Rufous elephant-shrew, *Elephantulus rufescens* (Macrocelididae), in Kenya. Z.Säugetierkunde 47: 65-71.

Neal, B.R. 1986. Reproductive characteristics of African small mammals. Cimbebasia (A) 8 (14): 113-127.

Neal, B.R. 1995. The ecology and reproduction of the short-snouted Elephant-shrew, *Elephantulus brachyrhynchus*, in Zimbabwe with a review of the reproductive ecology of the genus *Elephantulus*. Mammal Rev. 25: 51-60.

Nice, M.M. (1962). Development behavior in precocial birds. Transactions of the Linnean Soc. 8: 1-211.

Nicoll, M.E. and G.B. Rathbun. 1990. African Insectivora and Elephant-Shrews. An Action plan for their conservation. IUCN/SSC Insectivore, tree shrew and elephant-shrew specialist group. Gland, Switzerland.

Nikaido, N., Nishihara, H., Humoto, Y., and N. Okada. 2003. Ancient SINESs from African Endemic Mammals. Mol.Biol.Evol. 20: 522-527.

Nishihara, H., Y. Satta, M. Nikaido, J.G.M. Thewissen, M.J. Stanhope and N. Okada. 2006. A retroposon analysis of Afrotherian phylogeny. Mol. Biol. Evol. 22: 1823-1833.

Novacek.M.J. 1992. Mammalian phylogeny: shaking the tree. Nature 356: 121-125.

Novacek, M.J. 1984. Evolutionary stasis in the elephant shrew, *Rhynchocyon*. In: (N. Eldridge and S.M. Stanley, eds) Living fossils. Springer-Verlag, New York. pp. 4-22.

Nowak, R.M. and J.L Paradiso. 1983. Walker's Mammals of the World. John Hopkins Univ. Press, Baltimore, MD. ...

Nowak. R.M. 1991. Walker's Mammals ofthe World. Vol. 1. The John Hopkins University Press, Baltimore and London: 180-186.

Oduor-Okelo, D. Katema, R.M. and A. M. Carter. 2004. Placenta and fetal membranes of the four-toed elephant shrew, *Petrodromus tetradactylus*. Placenta 25: 803-809.

Olbricht, G., C. Kern and G. Vakhrusheva. 2005. Einige Aspekte der Fortpflanzungsbiologie von Kurzohr-Rüsselspringern (*Macroscelides proboscideus* A. Smith, 1829) in Zoologischen Gärten unter besonderer Berücksichtigung von Drillingswürfen. Zool. Garten N.F. 75: 304-316.

Olbricht, G. 2007. Longevity and fecundity in sengis (Macroscelidea). Afrotherian Conserv. 5: 3-5.

Olds, N. and J. Shoshani. 1982. *Procavia capensis*. Mammalian species 171: 1-7.

Partridge, L. 1989. Lifetime reproductive success and life-history evolution. In: (I. Newton, ed) Lifetime reproduction in birds. London.

Patterson, B. 1965. The fossil elephant shrews (Family Macroscelididae). Bull. Mus.Comp. Zool. 133.

Peaker, M. and E. Taylor. 1997. Sex ratio and litter size in the guinea pig. J. Reprod. Fert. 108: 63-67.

Pearl, R. 1913a. On the correlation between the number of mammae in the dam and the size of litter in mammals. I. Interracial correlation. Proc. Soc. for Exp. Biol. and Med. 10, Malden, Mass., USA: 27-30.

Pearl, R. 1913b. On the correlation between number of mammae of the dam and size of litter in mammals. II. Intraracial correlation in swine. Proc. Soc. Exp. Biol. and Med. 10, Malden, Mass., USA: 31-32.

Perrigo, G. 1987. Breeding and feeding strategies in deer mice and house mice when females are challenged to work for their food. Anim. Behav. 35: 1298-1316.

Perrin, M.R. 1986. Some perspectives on the reproductive tactics of southern African rodents. Cimbebasia (A) 8: 63-77.

Perrin, M. R. 1995a. The biology of elephant-shrews: Introduction. Mamm. Rev. 25: 1-2.

Perrin, M.R. (1995). Comparative aspects of the metabolismand thermal biology of elephant-shrews (Macroscelidea). Mamm. Rev. 25: 61-78.

Perrin, M.R. and L.J. Fielden. 1999. *Eremitalpa granti*. Mammalian species 629: 1-4.

REFERENCES

Peters, W.C.H. 1852. Naturwissenschaftliche Reise nach Mossambique. Georg Reimer, Berlin

Promislow, D.E. and P. H. Harvey. 1990. Living fast and dying young: a comparative analysis of life-history variation among mammals. J. Zool. 75: 224-226.

Pillay, N. 2001. Reproduction and postnatal development in the bush Karoo rat *Otomys unisulcatus* (Muridae, Otomyinae). J. Zool. 254: 515-520.

Pontier, D., J.M. Gaillard, D. Allainé, J. Trouvilliez, I. Gordon and P. Duncan. 1989. Postnatal growth rate and adult body weight in mammals: a new approach. Oecologia 80: 390-394.

Puschmann, W. 2004. Zootierhaltung. Säugetiere, Verlag Harri Deutsch Frankfurt/M., Germany.

Quay, W.B. 1977. Structure and function of skin glands. In: (D. Müller-Schwarze and M.M. Mozell, eds) Chemical signals in vertebrates, New York:Plenum Press.

Raman, J. and M.R. Perrin. 1997. Allozyme and isozyme variation in seven southern African elephant-shrew species. Z. Säugetierkunde 62: 108-116.

Rathbun, G.B. 1979a. The social structure and ecology of elephant-shrews. Z. Tierpsych. 20 (Suppl.): 1-77.

Rathbun, G.B. 1979b. *Rhynchocyon chrysopygus*. Mammalian species 117: 1-4.

Rathbun, G.B. 2009. Why is there discordant diversity insengi (Mammalia: Afrotheria. Macroscelidea) taxonomy and ecology?. Review article Afr. J. Ecol. 47: 1-13.

Rathbun G. B. and K. Redford. 1981. Pedal scent-marking in the rufous elephant-shrew. J. Mammal. 62: 635-637.

Rathbun, G.B., P. Beaman and E. Maliniak. 1981. Capture, husbandry and breeding of Rufous elephant-shrews, *Elephantulus rufescens*. Int. Zoo Yearbook 21: 176-184.

Rathbun, G.B. and P.F. Woodall. 2002. A bibliography of elephant-shrews or sengis (Macroscelidea). Mamm. Rev. 32: 66-70.

Rathbun G.B. and J. Kingdon. 2006. The etymology of "sengi". Afrotherian Conserv. 4: 14-15.

Rathbun, G.B. and C.D. Rathbun. 2006a. Social structure of the bushveld sengi (*Elephantulus intufi*) in Namibia and the evolution of monogamy in the Macroscelidea. J. Zool. Lond. (269): 391-399.

Rathbun, G.B. and C.D. Rathbun. 2006b. Social monogamy in the noki or dassie-rat (*Petromus typicus*) in Namibia. Z. Säugetierkunde/Mamm. Biol. 71: 203-213.

Rautenbach I.L. and D.A. Schlitter. Nongeographic variation in elephant shrews (Genus Elephantulus Thomas and Schwann, 1906) of Southern Africa. Annals of Carnegie Museum 46 (Article 14):223-243, 1977.

Renfree, M.B., E.S. Robinson, R.V. Short and J.L. Vanderberg. 1990. Mammary glands in male marsupials: I.Primordia in neonatal opossums *Didelphis virginiana* and *Monodelphis domestica*. Development 110: 385-390.

Reynolds III, J.E. and D.K. Odell. 1991. Manatees and Dugongs, Library of Congress Cataloguing-in-Publication Data, USA.

Ribble, D. 2003. The evolution of social and reproductive monogamy in *Peromyscus:* evidence from *Peromyscus californicus* (the California mouse). In: Monogamy - mating strategies and partnerships in birds, humans and other mammals. U. H. Reichard and C. Boesch (eds.), Cambridge University Press: 81-92.

Ribble, D.O. and M.R. Perrin. 2005. Social organization of the eastern rock elephant-shrew (*Elephantulus myurus*): the evidence for mate guarding. Belg. J. Zool. 135 (Suppl.): 167-173.

Roberts, A. 1951. The mammals of South Africa,:The "Mammmals of South Africa" Book Fund, Johannesburg.

Robinson, T. and E.P. Seiffert. 2004. Afrotherian origins and interrelationships: New views and future prospects. Current topics in developmental biology 63.

Romeis, B. 1989. Mikroskopische Technik, Urban and Schwarzenberg Verlag, München.

Romer, A.S. and T.S. Parsons. 1983. Vergleichende Anatomie der Wirbeltiere. Paul Parey Verlag, Hamburg und Berlin.

Rosenthal, M.A. 1975. The management, behavior and reproduction of the short-eared elephant shrew, Macroscelides proboscideus (Shaw). MS-Thesis, Dept. of Biology, Northeastern Univ. of Illinois.

Rovero, F., G.B. Rathbun, A. Perkin, T. Jones, D.O. Ribble, C. Leonard, R.R. Mwakisoma and N. Doggart. 2008. A new species of giant sengi or elephant-shrew (genus Rhynchocyon) highlights the exceptional biodiversity of the Udzungwa Mountains of Tanzania. J. Zool. 274: 126-133.

Roxburgh, L. and M.R. Perrin. 1994. Temperature regulation and activity pattern of the round-eared elephant shrew Macroscelides proboscideus. J. therm. biol. 19: 13-22.

Sacher G.A. and D.F. Staffeldt. 1974. Relation of gestation time to brain weight for placental mammals: implications for the theory of vertebrate growth. The Am. Natural. 108: 593-615.

Sauer, E.G.F. 1972. Setzdistanz und Mutterfamilie bei der Kurzohrigen Elefantenspitzmaus. Mitt. der Baseler Afrika Bibliographien (4-6): 114-134.

Sauer, E.G.F. 1973. Zum Sozialverhalten der Kurzohrigen Elefantenspitzmaus *Macroscelides proboscideus*. Zeitschrift für Säugetierkunde 38: 65-97.

Sauer, E.G.F. and E.M. Sauer. 1971. Die Kurzohrige Elefantenspitzmaus in der Namib. In: (Namib und Meer). Swakopmund, Südwest-Afrika: Ges. f. wiss. Entw. und Museum: 5-43.

Sauer, E.G.F. and E.M. Sauer. 1972. Zur Biologie der Kurzohrigen Elefantenspitzmaus. Z. des Kölner Zoos (15): 119-139.

Sánchez-Villagra, M.R., Y. Narita and S. Kuratani. 2007. Thoracolumbar vertebral number: the first skeletal synapomorphy for afrotherian mammals. Systematics and Biodiversity 5: 1-7.

Seiffert, E.R. 2002. The reality of afrotherian monophyly, and some of its implications for the evolution and conservation of Afro-Arabia's endemic placental mammals. Afrotherian Conserv. 1: 3-6.

REFERENCES

Seiffert, E.R. 2007. A new estimate of afrotherian phylogeny based on simultaneous analysis of genomic, morphological, and fossil evidence. BMC Evol. Biol. 7: 224.

Séguignes, M. 1989. Contribution à l'étude de la reproduction d'*Elephantulus rozeti* (Insectivora, Macroscelididae). Mammalia 53 (3): 377-386.

Sherman, P.W., S. Braude and J. Jarvis. 1999. Litter sizes and mammary numbers of naked mole-rats: breaking the one-half rule. J Mammalogy 80(3): 720-733.

Shigehara, N. 1980. Epiphyseal union, tooth eruption, and sexual maturation in the common tree shrew, with reference to its systematic problem. Primates 21: 1-19.

Shortridge, G.C. 1934. The mammals of South West Africa. Verlag Heinemann, London.

Shoshani, J., C.A. Goldman and J.G.M. Thewissen. 1988. *Orycteropus afer*. Mammalian species (No. 300): 1-8.

Simmons, N.B. 1993. Morphology, Function, and phylogenetic significance of pubic teats in bats (Mammalia: Chiroptera). A,. Mus. Novitates 3077: 1-37.

Simons, E. L., P.A. Holroyd and T.M. Brown. 1991. Early tertiary elephant-shrews from Egypt and the origins of the Macroscelidea. Proc. Natl. Acad. Sci. USA 88: 9734-9737.

Skinner, J.D. and H.N. Smithers. 1990. The Mammals of the Southern African Subregion, Univ. of Pretoria, South Africa.

Skinner J.D. and C.T. Chimimba. 2005. The Mammals of the Southern African Subregion, Cambridge Univ. Press, Cambridge.

Smithers, H.N. 1983. The Mammals of the Southern African Subregion, University of Pretoria, South Africa.

Sone, K., K. Koyasu, S. Kobayashi and S. Oda. Fetal growth and development of the coypu (Myocastor coypu): Prenatal growth, tooth eruption, and cranial ossification. Z. Säugetierkunde/Mamm.Biol. 73: 350-357.

Springer, M.S., G.C. Cleven, O. Madsen, W.W. de Jong, V.D. Waddell, H.M. Amrine and M.J. Stanhope. 1997. Endemic African mammals shake the phylogenetic tree. Nature 388: 61-64.

Springer, M.S., W.J. Murphy, E. Eizirik and S.J. O'Brien. 2003. Placental mammal diversification and the Cretaceous-Tertiary boundary. Proc. Natl. Acad. Sci. USA 100: 1056-1061.

Stanhope, M.J., M.R. Smith, V.G. Waddell, C.A. Porter, M.S. Shirji and M. Goodmann. 1996. Mammalian evolution and the interphotoreceptor retinoid binding protein (IRBP) gene: Convincing evidence for several superordinal clades. J. Mol. Evol. 43: 83-92.

Stanhope, M.J., O. Madsen, V.G. Waddell, G.C. Cleven, W.W. de Jong and M.S. Springer. 1998a. Highly congruent molecular support for a diverse superordinal clade of endemic African mammals. Mol. Phylogen. and Evol. 9: 501-508.

Stanhope, M.J., V.G. Waddell, O. Madsen, W.W. de Jong, SB. Hedges, G.C. Cleven, D. Kao and M.S. Springer. 1998b. Molecualr evidence for multiple origins of Insectivora and for a new order of endemic African insectivore mammals. Proc. Natl. Acad. Sci. USA 95: 9967-9972.

Starck, D. 1949. Ein Beitrag zur Kenntnis der Placentation bei den Macrosceliden. Z. Anat.Entw.gesch. 114: 319-339.

Stearns, S.C. 1992. The evolution of life histories. Oxford University Press London, UK.

Steiner, H.M. and J. Raczynski. 1976. Wiederholbarkeit von Messungen und individueller Messfehler bei craniometrischen Untersuchungen an Apodemus. Acta Theriol. 21: 535-541.

Stoch, Z.G. 1954. The male genital system and reproductive cycle of *Elephantulus myurus* jamesoni (Chubb). Phyl. Trans. R. Soc. Lond. 238: 99-126.

Stuart, C., T. Stuart and V. Pereboom. 2003. Aspects of the biology of the Cape Sengi, *Elephantulus edwardii*, from the Western Escarpment, South Africa. Afrotherian Conserv. 2:2-4.

Sumbera, R., H. Burda and W.N. Chitaukali. 2003. Reproductive biology of a solitary subterranean bathyergid rodent, the silvery mole-rat (*Heliophobus argenteocinereus*) J Mammology 84: 278-287.

Tabuce, R., R.J. Asher and T. Lehmann. 2008. Afrotherian mammals: a review of current data. Mammalia 72: 2-14.

Tolliver, D.K., L.W. Robbins, I.L. Rautenbach, D.A. Schlitter and C.G. Coetzee. 1989. Biochemical systematics of elephant shrews from southern Africa. Biochem. Syst. and Ecol. 17: 345-355.

Trillmich, F. 2000. Effects of low temperature and photoperiod on reproduction in the female wild guinea pig (*Cavia aprera*). J. Mammal. 81: 586-594.

Tripp, H.R.H. 1971. Reproduction in elephant-shrews (Macroscelididae) with special reference to ovulation and implantation. J. Reprod. Fertil. 26: 149-159.

Tripp, H.R.H. 1972. Capture, Laboratory care and breeding of elephant-shrews (Macroscelididae). Laboratory animals (6): 213-224.

Vakhrusheva, G. 2000. Reproductive behaviour of Short-eared elephant shrews. Mitt. BAG Kleinsäuger 2: 5-8.

van der Horst, C.J. 1942. The mechanism of egg transport from the ovary to the uterus in *Elephantulus*. S. Afr. J. Med. Sci. 8: 41-49.

van der Horst, C.J. 1944. Remarks on the systematics of *Elephantulus*. J. Mammal. 25: 77-82.

van der Horst, C.J. 1946. Some remarks on the biology of reproduction in the female of *Elephantulus*, the holy animal of Set. Trans. Roy. Soc. South Africa 31: 181-194.

van der Horst, C.J. and J. Gillman.. 1941a. The number of eggs and surviving embryos in *Elephantulus*. Anatom. Rec. 80: 443-452.

van der Horst, C.J. and J. Gillman. 1941b. The menstrual cycle in *Elephantulus*. S.Afr.J.Med.Sci. 6: 27-47.

van der Horst, C.J. and J. Gillman. 1942a. A critical analysis of the early gravid and pre-menstrual phenomena in the uterus of Elephantulus, Macaca and the human female. S. Afr. J. Med. Sci. 7: 134-143.

REFERENCES

van der Horst, C.J. and J. Gillman. 1942b. The life history of the corpus luteum of menstruation in Elephantulus. S. .Afr. J. Med. Sci. 7: 21-41.

van der Horst, C.J. and J. Gillman. 1942c. Pre-implantation phenomena in the uterus of Elephantulus. S. Afr. J. Med. Sci. 7: 47-71.

van der Horst, C.J. 1954. Elephantulus going into anoestrus; Menstruation and abortion. Phylosophical Transactions of the Royal Society of London.Series B, Biological Sciences 238 (653): 27-61.

van Dijk, M.A.M., O. Madsen, F. Catzeflis, M.J. Stanhope, W.W. de Jong and M. Pagel. 2001. Protein sequence signatures support the African clade of mammals. Proc. Natl. Acad. Sci. USA 98: 188-193.

Viljoen, S. and S.H.C. du Toit. 1985. Postnatal development and growth of southern African tree squirrels in the genera Funisciurus and Paraxerus. J. Mammal. 66: 119-127.

Waddell, P.J., H. Kishino and R. Ota. 2001. A phylogenetic foundation for comparative mammalian genomics. Genome Inform. 12: 141-154.

Ward, S.J. 1998. Numbers of teats and pre-and post-natal litter sizes in small diprotodont marsupials. J. Mammal. 79: 999-1008.

Weigl, R. 2005. Longevity of mammals in captivity; from the living collections of the world. Kleine Senckenberg-Reihe. Prof. F. Mosbrugger 48: 198-199.

Weir, B.J. 1971. Some notes on reproduction in the Patagonian mountain viscacha, Lagidium boxi (Mammalia: Rodentia). J. Zool. Lond. 164: 463-467.

Weir, B.J. and I.W. Rowlands. 1973. Reproductive strategies of mammals. Ann. Rev. Ecol. System. 4 (4056): 139-163.

Weir, B.J. 1974. Reproductive chcaracteristics of hystricomorph rodents. In (. I.W. Rowlands and B.J. Weir, eds) The biology of hystricomorph rodents. Zool. Soc. Lond. 34: 265-301.

Welsch, U., F. Feuerhake, R.J. van Aarde, W. Buchheim and S. Patton. 1998. A histo- and cytophysiology of the lactating mammary gland of the African elephant (*Loxodonta africana*). Cell Tissue Res. 292: 377-394.

Welsch, U. 2007. Secretory phenomena in the non-lactating human mammary gland. Ann. Anat. 189:131-141.

Wemmer, C., J. Watling, L. Collins and K. Lang. 1983. External characters of the Sulawesi palm civet, Macrogalidia musschenbroekii Schlegel, 1879. J. Mammal. 64: 133-136.

Werdelin, L. and A. Nilsonne. 1999. The evolution of the scrotum and testicular descent in mammals: a phylogenetic view. J Theoretical Biology 196: 61-72.

White, C.M.N. and W.F.H. Ansell. 1966. A list of Luvale and Lunda mammal names. The Puku 4: 181-185, 1966.

Wildman, D.E., C. Chen, O. Erez, L.I. Grossman, M. Goodmann and R. Romero. 2006. Evolution of the mammalian placenta revealed by phylogenetic analysis. Proc. Natl. Acad. Sci. USA 103: 3203-3208.

Withers, P.C. 1983. Seasonal reproduction by small mammals of the Namib desert. Mammalia 47: 195-204.

Woodall, P.F. 1995a. The male reproductive-system and the phylogeny of elephant-shrews (Macroscelidea). Mamm. Rev. 25: 87-93.

Woodall, P.F. 1995b. The penis of elephant shrews (Mammalia, Macroscelididae). J. Zool. Lond. 237: 399-410.

Woodall, P.F. and J.D. Skinner.1989. Seasonality of reproduction in male rock elephant shrews, *Elephantulus myurus*. J. Zool. Lond. (217): 203-212.

Woodall, P.F. and C. Fitzgibbon. 1995. Ultrastructure of spermatozoa of the yellow-rumped elephant shrew *Rhynchocyon chrysopygus* (Mammalia, Macroscelidea) and the phylogeny of elephant shrews. Acta Zool. 76: 19-23.

Yarnell, R.W. and D.M. Scott. 2006. Notes on the ecology of the short-snouted sengi (*Elephantulus brachyrhynchus*) at a game ranch in North-West Province, South Africa. Afrotherian Conserv. 4: 2-4.

Zeveloff, S.I. and S.M. Boyce. 1980. Parental investment and mating systems in mammals. Evolution 34 (5): 973-982.

Zeveloff, S.I. and S.M. Boyce. 1986. Maternal investment in mammals. Nature 321: 537-538.

Zullinger, E.M., R.E. Ricklefs, K.H. Redford and G.M. Mace. 1984. Fitting sigmoidal equations to mammal growth curves. J. Mammal. 65: 607-636.

VI APPENDIX

A LIST OF TABLES

Table 1.1 Life history characters of the Macroscelidea. — 12

Table 2.1 Measurements for penis location for individual male sengi specimens. — 35
Table 2.2 Measurements of penis location as means, SD and range of four different genera of Macroscelidea. — 36

Table 3.1. Measurements of teat placement in female sengis. — 47
Table 3.2 Horizontal and vertical location of teat pairs. — 48

Table 5.1 Growth parameter estimates. — 74
Table 5.2 Gompertz parameter ranking. — 75
Table 5.3 Differences in body parameters between male and female *M. proboscideus*. — 77

Table 6.1 Gompertz parameters for male and female *M. proboscideus*. — 88
Table 6.2 Means of asymptotic weight and growth constant. — 89

Table Appendix D. Raw data on individual body measurements of captive male and female *M. proboscideus*. — 112

B LIST OF FIGURES

Fig. 1.1 The four sengi genera. — 10

Fig. 2.1 Measurement methodology to determine penis location. — 33
Fig. 2.2 Penis location in *Macroscelides proboscideus*. — 34
Fig. 2.3 The Box and Whisker Plot illustrates the overlap in ranges among all four genera. — 36

Fig. 3.1 Measurement methodology to determine teat locations. — 43
Fig. 3.2 Male teats in *Petrodromus* and *Elephantulus rozeti*. — 46
Fig. 3.3 Two teat pairs in female *Rh. chrysopygus* and *P. tetradactylus*.— 45
Fig. 3.4. Three teat pairs in female *M. proboscideus* and *E. rupestris*. — 46
Fig. 3.5 Dorsolateral teat in a female rodent, *Petromus typicus*. — 49

Fig. 4.1 *Petrodromus tetradactylus*, female, overview. Azan stain.— 60
Fig. 4.2 *Petrodromus tetradactylus*, female, teat. HE, acini, teat canal and ducts. — 60
Fig. 4.3 *Petrodromus tetradactylus*, female, teat canal. Demonstration of actin. — 61
Fig. 4.4 *P. tetradactylus*, female, milk sinus and myoepithelial cells. Demonstration of actin. — 61
Fig. 4.5 *Petrodromus tetradactylus*, female, lining of acini. PNA. — 62
Fig. 4.6 *Petrodromus tetradactylus*, male, teat, acini and duct. HE. — 64
Fig. 4.7 *P. tetradactylus*, male, mammary gland. HE. — 64
Fig. 4.8 *Petrodromus tetradactylus*, male, mammary acini, WGA.— 65
Fig. 4.9 *P. tetradactylus*, male, scent glands. HE. — 66
Fig. 4.10 *P. tetradactylus*, female, scent glands. HPA. — 66

Fig. 5.1 Developmental stages of *M. proboscideus*. — 73
Figs. 5.2 - 5.8 Length of various body metrics of male and female *M. proboscideus* in relation to age. — 76
Figs. 5.9 and 5.10 Body length against hind foot length and body length against body mass. — 78

Fig 6.1 Data collection at Wuppertal Zoo. — 86
Fig. 6.2 Body mass increase for male and female *M. proboscideus*. — 87

C STAINING METHODS

Preparing the paraffin sections

After cooling and hardening of the paraffin blocks, a microtome was used for sections of 4µm. A paper strip helped to lay the individual sections into a water bath, where they could expand and thus, folds were smoothed away. After this the sections were separately put on a coated object slide and were firstly dried on a heat board and then with a temperature of 48 °C in an incubator over night.

Staining methods (Romeis 1989, Bancroft and Stevens 1990)

Haematoxylin and eosin stain (HE)

With this general staining cell nuclei are stained in blue, the cytoplasm and the muscle tissue in red. After dewaxing the sections were hydrated in descending alcohols (always five minutes in ethanol with concentrations of 100%, 96%, 80%, 70% and 50%). Then they were rinsed for five minutes in acid haemalum (1 g haematoxylin, 1000 ml Aqua dest., 200mg sodiumjodate, 50g potassium alum, 50g chloralhydrate and 1g citric acid), then firstly washed in aqua dest. and, finally under running tab water until sections "blue". For the next step the sections were first stained in 0.1% eosin solution for three minutes before rinsing with aqua dest. After dehydrating through ascending alcohol series (only briefly in 50%, 70%, 80%, 96% and 100% ethanol) and clearing 2x five minutes in xylol a mounting medium was used together with a cover glass to prepare the slide. At the end the section were left at room temperature for further drying.

Alcian blue stain (pH 2.5)

This method provides evidence for acid mucins which appear in bright blue. The procedure of dewaxing and hydrating through graded alcohol to water is the same as above. Then the sections are taken out of the Aqua dest. to be washed in 3% acetic acid for three minutes and finally in a solution of alcian blue (1g alcian blue 8GX in 100 ml 3% acetic acid). At the end the sections are rinsed in destilled water, passed through a dehydrating series of alcohol followed by xylene and then mounted.

APPENDIX STAINING METHODS

Periodic-Acid-Schiff technique (PAS)

With this technique aldehyde groups are formed thru oxidation of 1.2 glycol with periodic acid. The PAS-positive structures appear pink or purple whereas haematoxylin stains the nuclei blue.

As described above, the sections are dewaxed and brought to distilled water as above.. After this the sections oxidize with 1% periodic acid for seven minutes at room temperature. Then they are briefly rinsed in distilled water and covered with Schiff's solution (fuchsin sulphurous acid, purchased from Sigma, Deisenhofen, Germany, diluted 2:3) for three to five minutes. After that the sections are rinsed for one minute with sulphite water (200 ml aqua dest. with 10 ml 10% sodiumbisulphite and adding 10ml hydrochloric acid. After this they are cleared under running tab water for ten minutes before they are briefly stained with haematoxylin (se above). The sections are then stained blue again under running tab water, dehydrated through ascending alcohol series and xylene and eventually mounted.

Heidenhain's azan stain

In this stain method two acid dyes are used: azocarmine and aniline blue. A cytologic stain with azocarmine is combined with a counterstain with aniline blue, after mordanting with phosphotungstic acid. To obtain good results, it is necessary to overstain with azocarmine, then slowly differentiate with aniline-alcohol to prevent counterstain from covering stained muscle, collagen, etc.The nuclei will appear red, connective tissue blue and muscle tissue red to orange.

As described in the PAS technique, the sections are dewaxed and hydrated through graded alcohols to distilled water and cleared in xylene. Afterwards they are suspended for 10 to 15 minutes in warm azocarmine (0.1 g azocarmine in 100 ml Aqua dest.), briefely brought to the boil, cooled down and filtrated. Then watery acetic acid is added.

Afterwards the sections are rinsed thoroughly in Aqua dest. The differentiation of the connective tissue takes place shortly in aniline alcohol (0.1 ml aniline in 100 ml 96% ethanol). The differentiation is blocked in acetic acid alcohol (1ml acetic acid in 100 ml 96 % ethanol). Further differentiation and staining takes place for 30 minutes up to three hours in phosphowolframic acid solution (5 % watery solution) under microscope control. After that the sections are rinsed with Aqua dest. At the end they are treated with aniline blue-orange G (dilute 0.5 g aniline blue plus 2 g orange G in 100 ml Aqua dest., adding 8 ml acetic acid, boil up and filter). After a short rinse in Aqua dest. and ethanol (96 and 100 %), the sections are cleared in xylene and mounted.

APPENDIX STAINING METHODS

Histochemistry of lectins

The sections were first dewaxed and brought to water through descending alcohols, then treated for 10 minutes with 3% H_2O_2 to block the endogene peroxidase and eventually rinsed in PBS (phosphate buffered saline). For further blocking the sections are treated with 1% BSA (standard blocking solution) and subsequently placed in a humid chamber for 15 minutes. After that time the lectin in its adequate dilution is dripped thereon. After an incubation time of one hour in the humid chamber the material is rinsed in PBS buffer. The biotinylated lectins (purchased from Sigma, Deisenhofen, Germany) are applied to the sections, followed by ABC (Avidin-biotin peroxidase complexe, produced by Vector) and then incubated at room temperature for one hour. After a rinse in PBS again, DAB (diaminobenzidine dihydrochloride) is firstly activated by H_2O_2 and then incubated on the sections for 10 minutes. After a rinse with Aqua dest, the nuclei are counterstained slightly with hemalum. Then the tissue sections are dehydrated through ascending alcohols to xylol again, and mounted in DPX.

D TABLE

Raw data on body measurements of male and female captive born *Macroscelides proboscideus* with known birth and death dates: Weight, length of snout, ear, whisker, hindfoot, head/body and tail. The ID numbers identify their origin (Bitterwolf and Naumann are private breeders, Zool.Gardens Bernburg, Halle, Erfurt (especially marked) and Wuppertal)

ID number	Sex	Birth	Death	Death age (days)	Weight	Snout	Ear	Whisker	Tarsus	Head body	Tail
no ID	m	21.03.2003	21.03.2003	0	7,85	0,585	0,87		2,115	5,18	4,02
No ID	w	21.03.2003	21.03.2003	0	7,68	0,545	0,80		2,065	5,01	3,86
F20352D	w	02.09.2003	10.11.2003	69		1,21	2,13		3,11		8,31
Bitterwolf	m			0			0,92		2,13		3,65
Bitterwolf	w			0		1,04			2,33		4,42
M20352G	m	23.12.2003	11.01.2004	19		1,04	2,03	4,35	2,88	9,24	7,57
M203001X	m	06.06.2003	08.06.2003	2	6,99	0,59	1,05	2,31	2,24	5,95	
M203052H	m	27.12.2003	29.12.2003	2	8,65	0,66	1,11	2,68	2,44	6,46	5,06
M201012J	m	07.07.2001	09.07.2001	2		0,58	1,12	2,54	2,29	6,18	3,58
M94005B	m	18.01.1994	10.07.2001	2730		1,27	2,13	4,5	3,19	10,8	10,19
F2030015	w	30.07.2003	08.11.2003	101	33,23	0,91	2,38	5,51	3,14	11,59	8,94
Bitterwolf	w	28.03.2002	28.05.2003	426		1,04	2,45	4,88	3,24	11,59	9,78
203001S	u	19.05.2003	19.05.2003	0				2,1	2,13		
F203001B	w	05.01.2003	05.01.2003	0	8,4	0,6	1,17	2,16	2,15	6,61	3,57
F20301L	w	07.04.2003	07.08.2003	122		0,84	2,31	5,03	3,12	11,28	8,74
Bitterwolf	ID			0		0,44		2,34	2,23		4,07
Bitterwolf	w			0		0,47	1,19	2,35	2,18	6,41	4,05
F203052B	w	20.08.2003	20.08.2003	0		0,45	1,27	2,35	2,39		
M203052A	m	20.08.2003	20.08.2003	0		0,44	1,10	2,37	2,36		
M203001V	m	29.05.2003	01.06.2003	3		0,46	1,29	2,51	2,26	6,61	4,65
M203001Y	m	06.06.2003	06.06.2003	0		0,46	1,16	2,23	2,23	6,34	4,35
F203001W	w	06.06.2003	06.06.2003	0		0,45	1,31	2,24			
M2030016	m	04.08.2003	07.08.2003	3		0,48	1,11	2,14	2,14	6,06	4,45
M20004J	m	22.06.2000	02.04.2003	1014		1,13	2,44	5,19	3,24	9,72	9,49
F201012D	w	01.05.2001	05.06.2004	1131		1,11	2,21	5,06	3,17	11,1	10,61
M99021I	m	29.11.1999	05.10.2004	1772	38,98	1,21	2,25	4,82	3,08	11,97	

Naumann	w	02.02.2003	16.02.2003	14		0,84	1,51	3,99	2,58	7,69	6,32
Naumann	m	02.02.2003	16.02.2003	14		0,75	1,57	4,07	2,54	7,56	6,3
F204004Y	w	21.07.2004	23.07.2004	2		0,57	1,03				
F204004D	w	26.02.2004	01.08.2004	157		1,08	2,25	5,04	3,25	9,35	10,5
M206002N	m	12.06.2004	04.07.2004	22		0,94	1,25	5,03	2,91	8,46	6,59
M204004V	m	27.06.2004	28.06.2004	1	7,3	0,52	1,24	2,17	2,34	5,55	3,8
M2040041	m	08.08.2004	08.08.2004	0	8,05	0,54	1,13	2,17	2,24	6,52	4,69
F20404G	w	29.12.2004	16.02.2005	49	22,8	1,12	1,91	4,69	3,11	11,44	8,85
Fetus	w	04.06.2004	04.06.2004	0	4,79	0,37	0,67	1,44	1,68	5,27	2,78
Fetus	m	04.06.2004	04.06.2004	0	4,38	0,38	0,61	1,42	1,66	6,2	2,67
205007A	m	29.01.2005	29.01.2005	0	5,6	0,51	0,83	1,77	1,97	5,97	3,29
205007B	w	29.01.2005	29.01.2005	0	6,55	0,58	0,91	2,05	2,05	6,43	4,31
Fetus	m	24.05.2007	24.05.2007	0	2,13	0,44	0,43		1,03	3,37	1,54
F20202C	w	04.01.2002	25.01.2006	1482	40,33	1,16	2,02	5,44	3,29		
203230 Halle	w	20.07.2003	30.11.2004	499	48	1,25	2,19	5,6	3,18	9,6	12,7
1222 Bernb	m	21.07.2005	13.12.2005	145	29,7	0,98	2,14	5,3	3,06	9,6	7,3
M204004C	m	15.02.2004	04.01.2005	324	38,9	1,23	2,03	5,7	3,36	11,8	
M204004J	m	26.04.2004	18.03.2005	326	34	1,33	2,28	5,3	3,24	9,5	10,7
M2040048	m	07.10.2004	25.04.2005	200	35,52	1,08	2,02	5,4	2,87	10,3	10,07
F20507D	w	03.02.2005	05.05.2005	91		0,98	1,91	5,1	2,79	8,3	9,8
203299 Halle	m	23.08.2003	02.12.2003	101	32	1,14	2,12	5,5	3,29	9,2	9,24
M20507G	m	24.03.2005	01.05.2005	38	31,01		1,09	1,87	3,19	9,6	8,6
860 Bernb	m	24.02.2003	17.07.2004	509	37	1,16	2,45	5,3	3,25	8,65	9,75
F2040047	w	07.10.2004	20.04.2005	195	43,6	1,2	2,48	5,5	3,27	9,5	9,87
F20301L	w	07.04.2003	07.08.2003	122					3,07		8,44
F20602H	w	10.04.2006	14.12.2006	248	35,73	1,05	2,15	5,4	3,1	9,2	9,97
M206002R	m	04.08.2006	04.08.2006	0	6,27	0,523	0,59	1,77	2,01	5,4	3,14
M206002S	m	04.08.2006	04.08.2006	0	7	0,46	0,78	1,79	2,05	5,56	3,5
M20702I	m	17.03.2007	19.03.2007	2		0,49	1,01	2,58	2,19	4,92	4,13
205051 Halle	m	19.03.2005	05.04.2005	17		0,94	2,11	5,27	3,15	7,93	9,22
F20602AC	w	28.12.2006	30.12.2006	2		0,43	0,91		2,39	5,45	4,38
Erfurt 2812	m	10.02.2007	26.02.2007	16		0,6	1,53	4,37	2,71	6,03	6,53
1055 Bernb	w	21.06.2004	23.06.2004	2		0,42	0,79	2,23	2,11	3,82	3,4
F20404D	w	16.12.2004	16.12.2004	0		0,45	0,97	2,02	2,07	5	3,81
M202002A	m	02.01.2002	07.06.2007	1982		1,35	2,16	4,82	3,24	8,44	

M20702L	m	03.04.2007	06.04.2007	3		0,56	0,62	1,76	2,01	3,92	3,31
F20702M	w	02.04.2007	02.04.2007	0		0,5	0,88	2,29	2,1	4,1	3,64
2861 Erfurt	w	10.06.2007	29.06.2007	19		0,84	1,74	4,44	2,9	5,73	7,35
F20702Y	w	17.07.2007	21.07.2007	4	5,8	0,49	0,85	2,48	2,17	4,15	3,5
F20702X	w	17.07.2007	17.07.2007	0	7,28	0,5	0,69	2	2,24	4,8	3,53
F20507I	w	05.04.2005	24.05.2007	779		1,42	2,68	5,22	3,36	8,87	
F206002C	w	06.02.2006	24.02.2006	18		1,08	1,79	4,54	2,95	6,5	8,45
204067 Halle	m	30.03.2004	21.07.2004	113		1,01	2,26	5,26	3,2	9,2	9,67
M20602W	m	31.08.2006	01.10.2007	396		0,96	2,22	3,35	3,26	9,74	9,37

I want morebooks!

Buy your books fast and straightforward online - at one of the world's fastest growing online book stores! Environmentally sound due to Print-on-Demand technologies.

Buy your books online at
www.get-morebooks.com

Kaufen Sie Ihre Bücher schnell und unkompliziert online – auf einer der am schnellsten wachsenden Buchhandelsplattformen weltweit!
Dank Print-On-Demand umwelt- und ressourcenschonend produziert.

Bücher schneller online kaufen
www.morebooks.de

OmniScriptum Marketing DEU GmbH
Heinrich-Böcking-Str. 6-8
D - 66121 Saarbrücken
Telefax: +49 681 93 81 567-9

info@omniscriptum.com
www.omniscriptum.com

Printed by Books on Demand GmbH, Norderstedt / Germany